Lecture Notes in Mathematics 2176

More information about this series at http://www.springer.com/series/304

Anna M. Bigatti • Philippe Gimenez •
Eduardo Sáenz-de-Cabezón

Editors

Computations and Combinatorics in Commutative Algebra

EACA School, Valladolid 2013

 Springer

Editors

Anna M. Bigatti
Dipartimento di Matematica
Università degli Studi di Genova
Genova, Italy

Philippe Gimenez
Instituto de Investigación en Matemáticas
   (IMUVA)
Universidad de Valladolid
Valladolid, Spain

Eduardo Sáenz-de-Cabezón
Departamento de Matemáticas y
   Computación
Universidad de La Rioja
Logroño, Spain

ISSN 0075-8434   ISSN 1617-9692   (electronic)
Lecture Notes in Mathematics
ISBN 978-3-319-51318-8   ISBN 978-3-319-51319-5   (eBook)
DOI 10.1007/978-3-319-51319-5

Library of Congress Control Number: 2017934522

Mathematics Subject Classification (2010): 13-02, 13P99, 68W30, 05E40, 05E45

Printed on acid-free paper

This Springer imprint is published by Springer Nature
The registered company is Springer International Publishing AG
The registered company address is: Gewerbestrasse 11, 6330 Cham, Switzerland

Introduction

This is a book on *combinatorial commutative algebra*, more precisely, it describes some applications of the combinatorics of monomial ideals to more general results in commutative algebra. It covers three topics giving a good perspective of the different facets of this subject including recent developments as well as insights on some celebrated results. While each chapter can be read independently, the book as a whole has the extra value of providing a balanced overview of the richness of approaches to combinatorial commutative algebra.

Moreover, the volume has also a strong computational flavor. The three chapters contain computer sections, examples, and exercises for the reader. In the last years, very strong efforts have been made on the development of computational tools in commutative algebra and algebraic geometry, which have resulted in powerful computer algebra systems that are widely used by researchers. In this volume, chapter "Koszul Algebras and Computations" uses CoCoA [3], while chapters "Primary Decompositions" and "Combinatorics and Algebra of Geometric Subdivision Operations" use Macaulay2 [62]. Along the three chapters, we will assume that the basics on Gröbner bases and elimination contained in the celebrated book by Cox et al. [39] are known.

The first chapter is centered on Koszul algebras. This topic was first introduced in the noncommutative case in 1970; since then, it has been thoroughly studied revealing how it relates to many notions at the very heart of commutative algebra such as presentations, resolutions, Hilbert series, and complete intersections. This chapter explores the main results in these many aspects with the particular aim of showing how to investigate them computationally; thus, it gives detailed descriptions of the functions and techniques for creating new examples. It examines several types of Koszul algebras listing relevant classes of examples and gives precise references to the current literature on the topic. Finally, it provides a convenient table summarizing all definitions, implications, and examples in a way that can be very useful for the interested researcher.

The second chapter is devoted to primary decompositions, a topic that has historically been at the center of combinatorial commutative algebra and its computational aspect. Primary decompositions are a natural description of the structure of a module over a commutative ring. In the context of computational combinatorial commutative algebra, primary decompositions are a traditional topic. The first section of the chapter deals with the computation of primary decompositions and lists some of the many works this topic attracted with a special emphasis on the widely used algorithm of Gianni-Trager-Zacharias. The second section is focused on binomial ideals, which play a central role in the relation between commutative algebra and algebraic combinatorics. Moreover, binomial and toric ideals are of great importance in the relatively recent field of algebraic statistics. The third section of the chapter gives the reader a sample of algebraic statistics. The chapter finishes with yet another application of primary decompositions in the field of combinatorics, giving a short excursion into networks.

Finally, the last chapter of the book walks in the opposite direction of the path between algebra and combinatorics. Starting from abstract simplicial complexes that are combinatorial objects, the authors obtain significant results on algebraic structures associated to them. The main tool used in the chapter is the Stanley-Reisner ring of a simplicial complex, which turned out to be very fruitful since the early works of Stanley. The central topic of the chapter is the use of subdivision operators (which are of combinatorial nature) to find structural properties of algebraic objects. We find here again Koszul algebras. This demonstrates the mixed nature of the book and in fact of the area of combinatorial commutative algebra.

This book originated from a series of lectures at the EACA's Second International School on Computer Algebra and Applications. As it is a collection of lecture notes, each chapter of this volume includes examples and exercises that should help the reader to practice and make their own experiments on the topics presented here.

The EACA school [monica.unirioja.es/web_2EACA/index.html] was organized by the editors in the framework of the Spanish network RedEACA (*Red Temática de Cálculo Simbólico, Álgebra Computacional y Aplicaciones*), and it included, besides the series of lectures and their tutorial, some contributed talks by young participants. We would like to thank the invited speakers and all the participants and the members of the scientific committee of the RedEACA network for their help. Of course, we also want to thank the RedEACA network for providing the main financial support through the grant it received from the Spanish government (*Ministerio de Economía y Competitividad, grant MTM2011-13133-E*), the RSME (*Real Sociedad Matemática Española*) for providing some grants for young participants, and the IMUVA (*Instituto de Investigación en Matemáticas de la Universidad de Valladolid*) for hosting the EACA's school. Special thanks are due to the authors of the chapters who made an intense effort in the clarity of their exposition and were able to write a book that meets the highest standards from the scientific point of view and is, in our opinion, very representative of what computational combinatorial commutative algebra is nowadays.

Genova, Italy   Anna Maria Bigatti
Valladolid, Spain   Philippe Gimenez
Logroño, Spain   Eduardo Sáenz-de-Cabezón

Contents

Koszul Algebras and Computations

Anna M. Bigatti and Emanuela De Negri

Abstract A Koszul algebra R is a $\mathbb{N}$-graded K-algebra whose residue field K has a linear free resolution as an R-module. Many papers and lectures have been given on this topic, so here we collect various properties and facts which are related to being a Koszul algebra, and illustrate their mutual implications or counter-examples.

In addition we explain how one can investigate computationally these many aspects, some of which would seem to be intrinsically intractable, and we show many examples by using CoCoA-5.

1 Introduction

Koszul algebras are graded K-algebras R whose residue field K has a *linear* free resolution as an R-module. Koszul algebras were introduced by Priddy in the noncommutative case in 1970 [95], and have been recently studied from different point of views by many authors. We refer the reader to the recent surveys by Conca [34] and by Conca et al. [37] for the properties of Koszul algebras and their relation to regularity and other invariants, and to the volume by Polishchuk and Positselski [93] for various surprising aspects of Koszulness.

These notes collect the results presented in the course held in the EACA school by Aldo Conca. He gave a similar course in Levico and published his notes in *Koszul algebras and their syzygies* [34].

But here we present some of these results and examples from a different point of view: we emphasize on how to tackle these concepts with a computational approach by focusing on the specific theoretical results and on the practical techniques which are useful for dealing with examples and generating counterexamples. This approach is particularly interesting because many of the definitions and tools seem, at a first glance, to be practically intractable as intrinsically infinite or "generic". So we give complete and explicit examples using CoCoA-5; more specifically, we are using CoCoA-5.1.4, but should mostly work even in much newer versions. Most of

A.M. Bigatti (✉) • E. De Negri
Dipartimento di Matematica, Università degli Studi di Genova, Via Dodecaneso 35, 16146 Genova, Italy
e-mail: bigatti@dima.unige.it; denegri@dima.unige.it

© Springer International Publishing AG 2017
A.M. Bigatti et al. (eds.), *Computations and Combinatorics in Commutative Algebra*, Lecture Notes in Mathematics 2176, DOI 10.1007/978-3-319-51319-5_1

the functions we use are actually implemented in CoCoALib [2], the open source C++ mathematical heart of CoCoA.

For a more introductory approach to CoCoA-5 see also the chapter *Stanley decompositions using CoCoA* [17] (by the same authors) in the notes of another interesting summer school also devoted to the many applications of monomial ideals.

We start this chapter describing a well known technique: in Sect. 2 we show how to *present* a subalgebra R of $K[t_1, \ldots, t_r]$, i.e. write it as $R = K[x_1, \ldots, x_n]/I$ using *elimination*. Then we concentrate on subalgebras generated by monomials and compute their presentation using the optimized function "`toric`". Quite often these particular subalgebras exhibit all sorts of interesting behaviours, and are naturally easier to control and compute. We conclude the section listing some famous classes of such subalgebras.

We start our investigation in the context of standard graded algebras in Sect. 3 by defining *quadratic* and *G*-quadratic algebras. These are algebras whose ideal of the presentation is, for the first, an ideal generated by quadrics, and for the latter, an ideal with a Gröbner basis of quadrics (in some coordinate system). We also introduce the concept of *generic initial ideal* of an ideal and compare it being generated by quadrics with the other two properties on the ideal itself.

In Sect. 4 we define (and compute) resolutions, Castelnuovo-Mumford regularity, and describe the class of ideals *with 2-linear resolutions*, and compare this property with those of the previous sections.

At last in Sect. 5 we introduce the main stars of this course: *Koszul algebras*. Kosulness is quite hard to investigate computationally, since we have to deal with an infinite resolution. Again we find that there are implications on an algebra between being a Koszul algebra and the apparently unrelated conditions met before and also with arithmetic properties of its Hilbert Function.

Moreover we introduce another class of algebras: *LG-quadratic algebras* (Sect. 6). These are algebras that "Lift" to a G-quadratic algebra. This property is strictly weaker than being G-quadratic, but still implies Koszulness. Exploiting this concept computationally may be quite delicate and requires a good dose of knowledge and guesswork.

Conca et al. [36] introduced another interesting tool for proving Koszulness and generating Koszul algebras, the *Koszul Filtrations* (Sect. 7). If such a filtration exists for a standard graded algebra R then R is a Koszul algebras.

In Sect. 8 we talk about another class of examples which is quite interesting in this context: *complete intersections of quadrics*. We show that they rank between G-quadratic and LG-quadratic algebras. In this investigation we see how another computational technique, that is *solving polynomial systems*, either exactly or approximately, provides useful insights.

After touching so many interweaving different topics we conclude this course in Sect. 9 summarizing definitions, implications and examples into a compact table. We hope it will prove to be a valuable reference for your own exploration!

2 Presentation of Algebras

Let K be a field and $K[t_1, \ldots, t_r]$ a polynomial ring over K.

Given a set of n polynomials $N = \{f_1, \ldots, f_n\}$ in $K[t_1, \ldots, t_r]$ we consider the subalgebra of $K[t_1, \ldots, t_r]$ generated by N, denoted $K[N]$, and its presentation in the polynomial ring $K[x_1, \ldots, x_n]$:

$$\psi : K[x_1, \ldots, x_n] \longrightarrow K[N] \qquad x_i \mapsto f_i$$

Clearly $K[N] = K[x_1, \ldots, x_n]/\ker(\psi)$, and $\ker(\psi)$ is called **ideal of the presentation** of $K[N]$.

To calculate $\ker(\psi)$ one can use the well-known "elimination" procedure. We start by constructing the polynomial ring $K[t_1, \ldots, t_r, x_1, \ldots, x_n]$ and defining the ideal $I = (x_i - f_i \mid i \in 1, .., s)$. Then we have that $\ker(\psi) = I \cap K[x_1, \ldots, x_n]$.

Example 1 To present the algebra $\mathbb{Q}[t^3, t^4, t^5] \subset \mathbb{Q}[t]$ we need to compute the kernel of the homomorphism $\psi : \mathbb{Q}[x, y, z] \longrightarrow \mathbb{Q}[t^3, t^4, t^5]$ defined by $x \mapsto t^3$, $y \mapsto t^4$, $z \mapsto t^5$. Let's see how to compute it using CoCoA

```
/**/ use S ::= QQ[x,y,z, t];
/**/ I := ideal(x-t^3,  y-t^4,  z-t^5);
/**/ elim([t], I);
ideal(y^2 -x*z, -x^2*y +z^2, -x^3 +y*z)
```

CoCoA Remark. Handy Syntax for Defining a Polynomial Ring in CoCoA *The special syntax with "::=", or the command "use" (or "Use"), is interpreted as the mathematically familiar definition of a polynomial ring, such as "QQ[x,y,z]" for $\mathbb{Q}[x, y, z]$ and "QQ[x[1..n]]" for $\mathbb{Q}[x_1, .., x_n]$. Otherwise, if not preceeded by "::=" or "use", square brackets are interpreted as selector, for example "M[i,j]" means the entry the matrix M in position i, j.*

Elimination is a central topic in Computational Commutative Algebra (see for example the text books by Cox et al. [39] or by Kreuzer and Robbiano [69], Sect. 3.4) and its applications are countless.

Unfortunately this extraordinarily elegant tool often turns out to be quite inefficient, resulting in long and costly computations. Knowing how to exploit special properties of a given class of examples might make a huge difference.

And one very special case indeed is when the algebra we wish to present is generated by monomials: then the function "toric" uses a more efficient algorithm for the elimination and a dedicated syntax (see [18]).

Example 2 Again we consider $\mathbb{Q}[t^3, t^4, t^5]$:

```
/**/ use QQ[x,y,z];
/**/ toric(RowMat([3,4,5])); -- list of the exponents
ideal(-y^2 +x*z, x^3 -y*z, -x^2*y +z^2)
```

Example 3 With a very slightly more challenging example we can clearly measure the advantage in using "`toric`" over "`elim`":

```
/**/ use R ::= ZZ/(2)[x[1..6], s,t,u,v];
/**/ L := [s*u^20, s*u^30, s*t^20*v, t*v^20, s*t*u*v, s*t^2*u];
/**/ ExpL := [[ 1,  1,  1,  0, 1, 1],
              [ 0,  0, 20,  1, 1, 2],
              [20, 30,  0,  0, 1, 1],
              [ 0,  0,  1, 20, 1, 0]];

/**/ I := ideal([x[i] - L[i] | i in 1..6]);
/**/ t0 := CpuTime(); IE := elim([s,t,u,v], I); TimeFrom(t0);
9.274
/**/ t0 := CpuTime(); IT := toric(ExpL); TimeFrom(t0);
0.032

/**/ IT = IE;
true
```

CoCoA Remark *the CoCoA function "`toric`" follows a non-deterministic algorithm. The returned ideal is always the same ideal, but the set of generators of the ideal might vary.*

2.1 Some Families of Subalgebras Generated by Monomials

Even though subalgebras generated by monomials seem to be a very limited and special case, in practice they are often mentioned in examples because they present a great diversity of geometric properties with a simple description. Moreover their combinatorial nature provides many tools for their investigation.

Here we list some of the classical families of such subalgebras of the polynomial ring $K[t_1, \ldots, t_r]$.

Given $a = (a_1, \ldots, a_r) \in \mathbb{N}^r$ we denote by t^a the monomial $t_1^{a_1} \cdots t_r^{a_r}$ and by M_d the set of all monomials of degree d, that is $M_d = \{t^a \mid a \in \mathbb{N}^r,\ \deg(t^a) = d\}$.

1. **Veronese algebras**: fix $d \in \mathbb{N}$, take the algebra generated by M_d

$$V_{n,d} = K[M_d].$$

2. **Algebras of Veronese type**: fix $s = (s_1, \ldots, s_r) \in \mathbb{N}^r$ and take the algebra generated by the set $M_{s,d}$ of the monomial factors of t^s of degree d, that is $\{t^a \in M_d \mid t^a \text{ divides } t^s\}$ (Note: if $s = (d, \ldots, d)$, then $M_{s,d} = M_d$)

$$V_{s,d} = K[M_{s,d}].$$

3. **Pinched Veronese algebras**: fix $s \in \mathbb{N}$, and take the algebra generated by the monomials in M_d, supported in at most s indeterminates

$$\mathbf{PV}(r, d, s) = K[t^a \in M_d \mid \#\{i \mid a_i > 0\} \leq s].$$

4. Fix v in M_d, and take the algebra generated by $B(v)$, the smallest strongly stable subset (sometimes called *Borel subset*) of M_d containing v

$$B_v = K[B(v)].$$

5. Fix v in M_d, take the algebra generated by $B_s(v) = B(v) \cap M_{s,d}$

$$B_{s,v} = K[B_s(v)].$$

6. Fix $u, v \in M_d$, order M_d lexicographically, and consider the algebra generated by the *general lex-segment set* $L(u, v) = \{m \in M_d \mid u \geq m \geq v\}$

$$L_{u,v} = K[L(u, v)].$$

Example 4 The Pinched Veronese $\mathbf{PV}(3, 3, 2)$ is generated by all the monomials of degree 3 in x, y, z variables, except xyz

$$\{x^3, x^2y, x^2z, xy^2, xz^2, y^3, y^2z, yz^2, z^3\}$$

For the function "Toric" we write the list of exponents as columns, and define the polynomial ring with indeterminates $x_1, .., x_9$

```
/**/ M := mat([[3,2,2,1,1,0,0,0,0],
               [0,1,0,2,0,3,2,1,0],
               [0,0,1,0,2,0,1,2,3]]);
/**/ use QQ[x[1..9]];
/**/ toric(M);
```

```
ideal(-x[5]*x[6] +x[4]*x[8], -x[5]^2 +x[3]*x[9], -x[5]*x[7]+x[4]*x[9],
x[2]*x[5] -x[1]*x[8], x[3]*x[4] -x[1]*x[7], -x[3]*x[7] +x[2]*x[8],
x[3]*x[6] -x[2]*x[7], x[2]*x[4] -x[1]*x[6], -x[7]*x[8] +x[6]*x[9],
x[8]^2 -x[7]*x[9], -x[7]^2 +x[6]*x[8], -x[4]^2 +x[2]*x[6],
x[4]*x[5] -x[3]*x[7], x[3]^2 -x[1]*x[5], x[3]*x[5] -x[1]*x[9],
-x[3]*x[8] +x[2]*x[9], x[2]^2 -x[1]*x[4], x[1]*x[5]*x[6] -x[2]*x[3]*x[7],
x[5]^3 -x[1]*x[9]^2, x[5]^2*x[6] -x[3]*x[7]*x[8],
x[4]^3 -x[1]*x[6]^2, x[5]*x[6]^2 -x[4]*x[7]^2)
```

Recall that executing again "toric(M)" might return an ideal with a different list of generators (but the ideal is the same!): try calling "toric(M)" a few times and compare.

Example 5 When we want to test many examples making little changes to the paramenters, it is convenient to convert the CoCoA commands into a function. For example we show how to define the function for computing the B_v (Borel) algebras described above in point 3. First of all check in the manual if there is something about *strongly stable subsets*:

```
/**/   ? stable
```

this suggests to call "? StronglyStableIdeal" which gives this example:

```
------<  example  >-------
/**/   Use R ::= QQ[x,y,z];
/**/   L := [x*y^2*z];
/**/   StronglyStableIdeal(L);
ideal(x^4, x^3*y, x^2*y^2, x*y^3, x^3*z, x^2*y*z, x*y^2*z)
```

whose list of generators is exactly what we need! So here is how we can compute the ideal of the presentation of $B(v)$.

CoCoA Remark *the CoCoA function "*exponents*" was called "*log*" in older versions.*

```
define B(v)
  N := gens(StronglyStableIdeal([v]));
  M := transposed(mat([ exponents(pp) | pp in N ]));
  K := CoeffRing(RingOf(v));
  R ::= K[x[1..NumCols(M)]]; // create a new ring with K
  return toric(R, M);
enddefine;
```

then we can call it as any predefined CoCoA function:

```
/**/ use QQ[t[1..3]];
/**/ B(t[2]^3*t[3]);
```

```
ideal(-x[2]*x[3] +x[1]*x[4], -x[3]*x[8] +x[2]*x[9], -x[3]^2 +x[1]*x[5],
-x[7]^2 +x[6]*x[8], -x[4]*x[7] +x[3]*x[8], -x[3]*x[7] +x[1]*x[9],
x[8]^2 -x[7]*x[9], x[2]*x[6] -x[1]*x[7], -x[5]*x[7] +x[3]*x[9],
-x[5]*x[8] +x[4]*x[9], -x[3]^2 +x[2]*x[4], x[4]*x[6] -x[3]*x[7],
-x[7]*x[8] +x[6]*x[9], -x[5]*x[7] +x[4]*x[8], x[2]^2 -x[1]*x[3],
-x[2]*x[7] +x[1]*x[8], -x[3]*x[4] +x[2]*x[5], -x[3]*x[7] +x[2]*x[8],
x[3]*x[6] -x[2]*x[7], x[5]*x[6] -x[4]*x[7], x[4]^2 -x[3]*x[5])
```

CoCoA Remark. Generators vs. Minimal Generators *If not explicitely requested, a set of generators is not minimal. The operation of minimalizing ("MinGens", "MinSubsetOfGens", "minimalized") might be very expensive, so it is not performed by default. If you want "B(v)" to be minimally generated just substitute the line "return toric(R, M);" with "return minimalized(toric(R, M));"*

Example 6 The Example 4 was small enough to write the generators of **PV**$(3, 3, 2)$ "by hand", but now we define a function to write them for us, creating automatically a suitable polynomial ring (the CoCoA function "NewPolyRing" is the most flexible way to create a polynomial ring):

```
define PinchedVeroneseGens(r,d,s)
  R := NewPolyRing(RingQQ(), SymbolRange("t",1,r));
  Md := support(sum(indets(R))^d); -- all monomials of degree d
  L := [ T in Md |
         len([e in exponents(T) | not(IsZero(e))]) <= s ];
  return L;
enddefine;

/**/ PinchedVeroneseGens(3,3,2);
[t[1]^3, t[1]^2*t[2], t[1]*t[2]^2, t[2]^3, t[1]^2*t[3],
 t[2]^2*t[3], t[1]*t[3]^2, t[2]*t[3]^2, t[3]^3]
```

(Note that this list of generators is a permutation of the one we wrote "by hand" in Example 4.)

Even more useful for what we will want to compute, we can also define a function returns directly the ideal of the presentation of $\mathbf{PV}(r,d,s)$, minimally generated, by calling "PinchedVeroneseGens":

```
define PinchedVeroneseIdeal(r,d,s)
  PVGens := PinchedVeroneseGens(r,d,s);
  M := transposed(mat([exponents(T) | T in PVGens]));
  R := NewPolyRing(RingQQ(), SymbolRange("x",1,NumCols(M)));
  return minimalized(toric(R, M));
enddefine;

/**/ PinchedVeroneseIdeal(3,3,2);

ideal(x[8]^2 -x[6]*x[9], -x[6]*x[8] +x[4]*x[9], -x[5]*x[8] +x[2]*x[9],
-x[7]^2 +x[5]*x[9], -x[6]*x[7] +x[3]*x[9], x[5]*x[7] -x[1]*x[9],
-x[4]*x[7] +x[3]*x[8], x[2]*x[7] -x[1]*x[8], -x[6]^2 +x[4]*x[8],
-x[5]*x[6] +x[2]*x[8], x[3]*x[7] -x[2]*x[8], x[5]^2 -x[1]*x[7],
x[4]*x[5] -x[2]*x[6], x[3]*x[5] -x[1]*x[6], -x[3]^2 +x[2]*x[4],
x[2]*x[3] -x[1]*x[4], x[2]^2 -x[1]*x[3] )
```

3   Quadratic and G-Quadratic Algebras

We notice that the ideal in the last example is generated by quadrics (i.e. homogeneous polynomials of degree 2). In this section we aim our attention at two key properties related with sets of quadrics.

Let K be a field, and let $S = K[x_1,\ldots,x_n]$ be standard graded, i.e. $\deg(x_i) = 1$ for all x_i.

Definition 7 An algebra is **quadratic** if its defining ideal has a generating set made of quadrics.

Example 8 The Pinched Veronese algebra $\mathbf{PV}(3,3,2)$ is quadratic (recall Example 6, and the CoCoA functions defined there), but not all the Pinched Veronese algebras are quadratic: $\mathbf{PV}(4, 5, 2)$ is minimally defined by 168 quadrics and 12 cubics.

```
/**/ L := PinchedVeroneseIdeal(4,5,2);
/**/ len([ g in gens(L) | deg(g) = 2]);
168
/**/ len([ g in gens(L) | deg(g) = 3]);
12
/**/ len([ g in gens(L) | deg(g) > 3]);
0
```

Definition 9 An algebra is **G-quadratic** if its defining ideal I has a Gröbner basis (with respect to some coordinate system and some term order) made of quadrics. In other words, $R = S/I$ is G-quadratic if there exists a K-basis of S_1, say $x_1, \ldots, x_n$ and a term order σ such that the initial ideal of I with respect to σ, $\mathrm{in}_\sigma(I)$ (or $\mathrm{LT}_\sigma(I)$), is generated by monomials of degree 2 in $x_1, \ldots, x_n$.

CoCoA Remark *Often* in_σ *is also called* **leading term** *and is denoted* LT_σ. *This is the name used in CoCoA.*

It follows immediately that if an algebra is G-quadratic then it is quadratic.

It is well known ([9] by Backelin, [10] by Backelin and Fröberg) that the Veronese algebras V_d are G-quadratic. In [110, Theorem 14.2], Sturmfels proves that also the algebras of Veronese type, $V_{s,d}$, are G-quadratic, by using essentially an elimination argument. In [41] it is proved that $B_{s,v}$ and $L_{u,v}$ are also G-quadratic.

Example 10 Recall the function definition for Example 5: we can compute a minimal set of generators ("MinGens") or a Gröbner basis ("GBasis") for "B(t[2]^3*t[3])" and check their degrees. Even easier, we can ask CoCoA to print all the degrees, or just the maximum degree in a list:

```
/**/ KerPsi := B(t[2]^3*t[3]);
/**/ MinG := MinGens(KerPsi);
/**/ [ deg(g) | g in MinG ];
[2, 2, 2, 2, 2, 2, 2, 2, 2, 2, 2, 2, 2, 2, 2, 2, 2, 2, 2]
/**/ max( [ deg(g) | g in MinG ] );
2
/**/ max( [ deg(g) | g in GBasis(KerPsi) ] );
2
```

So we showed that $B(t_2^3 t_3)$ is quadratic and also G-quadratic.

Example 11 Now we refine the definition of the function "B" from Example 5 to generate the $B_{s,v}$ algebras described in point 4 of Sect. 2.1. We need just a tiny change:

```
define Bs(s,v)
  N := gens(StronglyStableIdeal([v]));
  M := transposed(mat([ exponents(f) |
                            f in N and IsDivisible(f,s) ]));
  K := CoeffRing(RingOf(v));
```

```
   R ::= K[x[1..NumCols(M)]];
   return toric(R, M);
enddefine;
```

and then we can test this class of examples:

```
/**/ use QQ[t[1..3]];
/**/ KerPsi := Bs(t[2]^2, t[2]^3*t[3]);   indent(KerPsi);
ideal(
  -x[2]*x[4] +x[1]*x[5],
   x[3]*x[4] -x[2]*x[5],
  -x[2]^2 +x[1]*x[3]
)
/**/ max( [ deg(g) | g in MinGens(KerPsi) ] );
2
/**/ max( [ deg(g) | g in GBasis(KerPsi) ] );
2
```

So we showed that $B_{(0,2,0)}(t_2^3 t_3)$ is quadratic and also G-quadratic.

From the definition we have that a G-quadratic algebra has a Gröbner basis made of quadrics in *some* coordinate system with *some* term order. And this clearly poses a big problem!

Example 12 Consider again the Pinched Veronese **PV**$(3, 3, 2)$. As we have shown in Example 4 this algebra is quadratic:

```
/**/ J := PinchedVeroneseIdeal(3,3,2);
/**/ max([deg(f) | f In gens(J)]);
2
```

Is it G-quadratic? it is not known!

The typical approach is to start by checking the reduced Gröbner basis for some orders. Indeed there are non-quadratic elements.

We start calculating the Gröbner basis of the ideal w.r.t. the order we have (the default in CoCoA is StdDegRevLex):

```
/**/ [deg(f) | f In GBasis(J)];
```

```
[2, 2, 2, 2, 2, 2, 2, 2, 2, 2, 2, 2, 2, 2, 2, 2, 2, 3]
```

Then we try changing term order, for example lex:

CoCoA Remark *The term order is an intrinsic attribute of a polynomial ring P, so if we want to try with other term order we need to define a new ring R, and the homomorphism from P into R sending $x_i \mapsto x_i$.*

```
/**/ R ::= QQ[x[1..9]], lex; -- no need of command "use"
/**/ phi := PolyAlgebraHom(RingOf(J), R, indets(R));
```

```
/**/ JLex := ideal(apply(phi, gens(J))); -- ideal in R
/**/ [deg(f) | f In GBasis(JLex)];
```

```
[2, 2, 2, 2, 2, 2, 2, 2, 2, 2, 2, 2, 2, 2, 2, 2, 2, 3, 3, 3, 3,
    3, 3]
```

CoCoA Remark. Ring of an Ideal vs. Current Ring *An ideal I generated by polynomials in a ring R is always seen as an object in R, so its Gröbner basis is computed in its ring R whatever is the current ring (i.e. the one introduced with the command "use") at the moment of the function call "GBasis(I)".*

We can also define a polynomial ring whose order is defined by a matrix (with the function "NewPolyRing"). For example we can try with some random weights: we create an order matrix *M* and we specify that the *grading dimension* is 1 (given by the first 1 row(s) of *M*):

```
/**/ W := RowMat([random(1,5) | i in 1..9]);  W;
matrix(QQ,
 [[4, 1, 2, 4, 1, 3, 4, 5, 4]])
/**/ M := MakeTermOrd(W);
/**/ R := NewPolyRing(QQ, SymbolRange("x",1,9), M, 1);
/**/ phi := PolyAlgebraHom(RingOf(J), R, indets(R));

/**/ JSigma := ideal(apply(phi, gens(J)));
/**/ [deg(f) | f in GBasis(JSigma)];
```

```
[2, 2, 3, 2, 2, 2, 2, 3, 2, 2, 2, 2, 2, 2, 2, 2, 3, 2, 2, 3, 3,
    2, 3]
```

CoCoA Remark *The function "MakeTermOrd" was called "CompleteToOrd" in older versions.*

CoCoA Remark. Degree vs. Weighted-Degree *The function "deg" always returns the standard degree. The function for the weighted-degree is "wdeg" and returns a list whose length is the grading dimension, in this case 1:*

```
/**/ [wdeg(f) | f In GBasis(JSigma)];

[[5], [5], [5], [6], [6], [6], [6], [6], [7], [7], [8], [8],
[8], [8], [8], [8], [8], [9], [9], [9], [6], [10], [12]]
```

In fact, there are finitely many orders giving different Gröbner bases (see about the *Gröbner Fan* in [79] by Mora and Robbiano) and we can determine all of them ([66, 67] by Jensen). So we can compute all the reduced Gröbner bases for *J* and verify they are not quadratic.

```
/**/ GBs := GroebnerFanIdeals(J);
```

For the specific example above, the problem is not time, but space, because the function actually stores all the Gröbner bases, and quickly takes more that 4Gb of RAM. Then we had to use a more technical function which calls a specified function on each Gröbner basis as soon as it is computed, and without storing it. For example we asked to print the maximum degree:

```
define PrintMaxDegInGB(I)
  print max([deg(g) | g in ReducedGBasis(I)]);
enddefine;
```

and then we got the answer using only 17Mb of RAM (in a few hours) by calling

```
/**/ CallOnGroebnerFanIdeals(J, PrintMaxDegInGB);
```

```
*3*3*3*3*3*3*3*3*4*4*4*3*3*3*5*7*7*7*7*7*7*7*7*7*7*7*7*7*5
*3*3........
```

where each * is a GBasis computation, and the numbers are the output of our PrintMaxDegInGB.

In conclusion, we have computed that there are 54828 different Gröbner bases (no wonder it takes time!!), and there is no order σ such that the reduced σ-Gröbner basis is made of quadrics. However, it is not known whether in other coordinates there is a quadratic Gröbner basis.

CoCoA Remark. CoCoA and GFan *From version CoCoA-5.1.3, thanks to Anders Jensen, CoCoA can access the library* gfanlib. *This uses arbitrary precision integers so it is slower than* gfan *itself, which uses machine integers.*

Notice that CoCoA doesn't store the Gröbner bases as such, but stores the ideals: an ideal knows its polynomial ring, therefore the ordering, and its Gröbner basis, and it keeps this informations in a more complete and usable form than just a bare list of polynomials.

Open question: for which values of n, d, s is the algebra $\mathbf{PV}(n, d, s)$ quadratic or G-quadratic?

Exercise 13 We showed that $PV(4, 5, 2)$ is not quadratic. Find other examples.

Exercise 14 Experiment with the function "PinchedVeroneseIdeal" to find classes of quadratic and non-quadratic algebras in this family.

3.1 Generic Initial Ideals

Another interesting tool in this context is the **generic initial ideal**, denoted $\text{gin}_\sigma(I)$. It is defined as follows: given an ideal I in $K[x_1, \ldots, x_n]$ and an order σ. Consider a generic change of coordinates g (i.e. $g(x_j) = \sum_{i=1}^n a_{i,j} x_i$ in $K(a_{i,j})[x_1, \ldots, x_n]$) then $\text{gin}_\sigma(I) = \text{in}_\sigma(g(I))$.

Example 15 If K is a small finite field it may happen that there is no actual change of coordinates h such that $\text{gin}_\sigma(I) = \text{in}_\sigma(h(I))$.

For instance consider the principal ideal $I = (x \cdot y \cdot (x + y)) \subset \mathbb{Z}/(2)[x, y]$ with order lex, and the generic change of coordinates $g(x) = ax + by$, $g(y) = cx + dy$; hence $g(x \cdot y \cdot (x + y)) = a \cdot c \cdot (a + c)x^3 + \dots$ and then $\mathrm{gin}(I) = (x^3)$.

This is one way to do this computation in CoCoA: first we make $K = \mathbb{Z}/(2)(a, b, c, d)$, and then we define $g : K[x, y] \longrightarrow K[x, y]$

```
/**/ K := NewFractionField(NewPolyRing(NewRingFp(2),"a,b,c,d"));
/**/ use R ::= K[x,y];

/**/ g := PolyAlgebraHom(R, R, [a*x+b*y, c*x+d*y]);
/**/ g(x*y*(x+y));
(a^2*c +a*c^2)*x^3 +(b*c^2 +a^2*d)*x^2*y +...
```

But for any specific values of a and c in $\mathbb{Z}/(2)$ we have $a \cdot c \cdot (a + c) = 0$ so for no actual change of coordinates $h : \mathbb{Z}/(2)[x, y] \longrightarrow \mathbb{Z}/(2)[x, y]$ we can get $\mathrm{in}(h(x \cdot y \cdot (x + y))) = x^3$:

```
/**/ h := PolyAlgebraHom(R, R, [x, x+y]);
/**/ h(x*y*(x+y));
x^2*y +x*y^2

/**/ h := PolyAlgebraHom(R, R, [x+y, y]);
/**/ h(x*y*(x+y));
x^2*y +x*y^2
```

On the other hand, if K is an infinite field there are infinitely many changes of coordinates h in $K[x_1, \dots, x_n]$ such that $\mathrm{gin}_\sigma(I) = \mathrm{in}_\sigma(h(I))$, more precisely, they form a Zariski open set. So in this case $\mathrm{gin}_\sigma(I)$ can be computed with high probability using a *random* (instead of generic) change of coordinates.

Example 16 Consider the example above with $K = \mathbb{Q}$: the principal ideal $I = (x \cdot y \cdot (x + y)) \subset \mathbb{Q}[x, y]$ with order lex, and again we have $\mathrm{gin}(I) = (x^3)$.

```
/**/ K := QQ;
/**/ use R ::= K[x,y];

/**/ h := PolyAlgebraHom(R, R, [2*x-5*y, -2*x+7*y]);
/**/ h(x*y*(x+y)); --> not good
-8*x^2*y +48*x*y^2 -70*y^3

/**/ h := PolyAlgebraHom(R, R, [7*x-7*y, -2*x+7*y]);
/**/ h(x*y*(x+y)); --> good
-70*x^3 +315*x^2*y -245*x*y^2

/**/ h := PolyAlgebraHom(R, R, [2234*x-5325*y, 1051*x+7515*y]);
/**/ h(x*y*(x+y)); --> good
7712963190*x^3 +41907481935*x^2*y -106946739225*x*y^2 .....
```

Now, with this new tool, we go back to our investigation on G-quadratic algebras.

Remark 17 If K is infinite and I is such that $\text{gin}_\sigma(I)$ is generated by monomials of degree 2 for some term order σ, then R/I is G-quadratic: in fact there exists a change of coordinates h (indeed, lots of them: a Zariski open set!) such that $\text{in}_\sigma(h(I))$ is generated in degree 2; that means that the σ-Gröbner basis of $h(I)$ is made of quadrics, and therefore R/I is G-quadratic.

Example 18 Consider the Borel algebra "B(t[2]^3*t[3])". We have seen in Example 10 that it is G-quadratic (in the given coordinates). Now we compute its gin (reminder: the CoCoA definition of the function "B(v)" is in Example 5).

```
/**/ use S ::= QQ[t[1..3]];
/**/ I := B(t[2]^3*t[3]);
/**/ gin(I);
ideal(x[1]^2, x[1]*x[2], x[2]^2, x[1]*x[3], x[2]*x[3], x[3]^2,
x[1]*x[4], x[2]*x[4], x[3]*x[4], x[4]^2, x[1]*x[5], x[2]*x[5],
x[3]*x[5], x[4]*x[5], x[5]^2, x[1]*x[6], x[2]*x[6], x[3]*x[6],
x[4]*x[6], x[5]*x[6], x[6]^2)
```

So we also have that $\text{gin}(I)$ is generated in degree 2.

CoCoA Remark. How Random Is "Random"? *Usually, when choosing random coefficients, one needs to strike a balance between a very big range, to provide a behaviour which is "generic enough", but not too big, to limit the coefficient growth during the computation. The implementation for gin in CoCoA uses a special representation for rational coefficients,* NewRingTwinFloat, *given as pairs of floating-point numbers with control on the propagation of the rounding errors, and this allows the random choice for the coefficients to be in the range* $(-10^6, 10^6)$.

Example 19 We present two examples introduced by Roos in [99] (numbers 22 and 72) of G-quadratic algebras with non-quadratic gin.

CoCoA Remark. Invisible Multiplication *CoCoA-5 can accept the convenient old CoCoA-4 syntax without multiplication signs for expressions written between two triple-stars "***" as in the definition of "*I22*" below, but then one needs to use the CoCoA-4 syntax rules: indeterminates must be lower-case letters, and functions must be capitalized.*

```
/**/ Use T ::= QQ[x,y,z,u];
/**/ I22 := *** Ideal(xz, y^2, z^2, yu + zu) ***;
/**/ LT(I22);
ideal(y*u, z^2, x*z, y^2)
```

So the generators of I_{22} form a Gröbner basis, and therefore I_{22} defines a G-quadratic algebra, but the $\text{gin}(I_{22})$ is not generated by quadrics:

```
/**/ gin(I22);
ideal(x^2, x*y, y^2, x*z, y*z^2, y*z*u)
```

The following ideal is even more special, being generated by quadratic monomials, thus it naturally defines a G-quadratic algebra, and again $\text{gin}(I_{72})$ is not generated by quadrics:

```
/**/ I72 := ***Ideal(xz, y^2, yu, z^2, zu, u^2)***;
/**/ gin(I72);
ideal(x^2, x*y, y^2, x*z, y*z, z^2, x*u^2)
```

4 Resolutions

Resolutions are the tool for defining Koszul algebras, and we first introduce them here in the setting of polynomial rings.

Definition 20 Let K be a field, and let $S = K[x_1,\dots,x_n]$ be standard graded.

Any finitely generated graded S-module $M = \oplus_{i\in\mathbb{Z}}M_i$ has a finite minimal graded resolution

$$0 \to \bigoplus_j S(-j)^{\beta_{pj}(M)} \to \cdots \to \bigoplus_j S(-j)^{\beta_{0j}(M)} \to M \to 0.$$

The numbers $\beta_{ij}(M)$ are called the **graded Betti numbers** of M, and $\beta_i(M) = \sum_j \beta_{ij}(M)$ the ordinary **Betti numbers** of M.

The resolution is known to have length $p \le n$.

Example 21 Given a homogeneous ideal I we can compute its minimal resolution

CoCoA Remark *CoCoA always prints the ring in a resolution using the letter "R".*

```
/**/ use S ::= QQ[x,y,z];
/**/ I := toric(S, [2,3,4]);  I;

/**/ I := ideal(x^2*y +z^3, y^2 -x*z, y*z-z^2);
/**/ PrintRes(I);
0 --> R(-6)^2 --> R(-4)(+)R(-5)^3 --> R(-2)^2(+)R(-3)
/**/ PrintBettiMatrix(I);

--    -->  -->   --
    0    0    0
    0    0    2
    0    0    1
    0    1    0
    0    3    0
    2    0    0
--    -->  -->   --
```

CoCoA Remark. Betti Matrix vs. Betti Diagram *Traditionally the resolution is written from right to left and the CoCoA function "PrintBettiMatrix" reflects that style in printing right to left the graded Betti numbers.*

*A more compact and well known way of printing the Betti numbers is the **Betti diagram**: the i-th Betti numbers $\beta_{ij}(M)$ are shifted up by i, and they are written from left to right, according to their indices:*

```
/**/ PrintBettiDiagram(I);

            0    1    2
     --------------------
     2:     2    -    -
     3:     1    1    -
     4:     -    3    2
     --------------------
    Tot:    3    4    2
```

We introduce now an important invariant of M that measure the growth of the Betti numbers of M.

Definition 22 The **Castelnuovo-Mumford regularity** of M is defined as:

$$\text{reg}(M) \;=\; \sup\{j - i \mid \beta_{ij}(M) \neq 0\}.$$

Note that the Castelnuovo-Mumford regularity is effectively the "deepest" level in the Betti diagram.

4.1 Ideals with 2-Linear Resolution

We introduce now a class of ideals, which in particular define quadratic and G-quadratic algebras.

Let M be a graded S-module, and $\beta_{ij}(M)$ its graded Betti numbers.

Definition 23 We say that M has a *d*-**linear resolution** if $\beta_{ij}(M) = 0$ for every $i \neq j + d$.

In particular if M has a d-linear resolution, then M is generated in degree d, and its resolution is of the form:

$$0 \to S(-d-p)^{\beta_p(M)} \to \cdots S(-d-1)^{\beta_1(M)} \to S(-d)^{\beta_0(M)} \to M \to 0$$

It is clear that M has a d-linear resolution if and only if $\text{reg}(M) = d$.

Example 24 The ideal $I = (wxy^2, x^2y^2, x^2yz, xy^2z)$ has a 4-linear resolution (and therefore its Betti Diagram prints in just one line).

```
/**/ use S ::= QQ[w,x,y,z];
/**/ I := ***Ideal(w*x*y^2, x^2*y^2, x^2*y*z, x*y^2*z)***;
/**/ PrintRes(I);
0 --> R(-6) --> R(-5)^4 --> R(-4)^4

/**/ PrintBettiDiagram(I);
```

```
              0     1     2
    ---------------------------
    4:    .  4     4     1
    ---------------------------
    Tot:     4     4     1
```

Now consider a homogeneous ideal I in $S = K[x_1, \ldots, x_n]$. As we have noticed before, if I has a 2-linear resolution, then I is generated in degree 2, that is, S/I is quadratic. But having a 2-linear resolution (equivalently to having regularity 2) is a stronger property.

Proposition 25 *If I is a homogeneous ideal in $S = K[x_1, \ldots, x_n]$ (with K infinite) whose resolution in 2-linear resolution then* $\mathrm{gin}_{\mathrm{DegRevLex}}(I)$ *is generated in degree 2 (and therefore R/I is G-quadratic by Remark 17).*

Proof In general in(I) has higher regularity than I, but Bayer and Stillman [12] proved that $\mathrm{reg}(I) = \mathrm{reg}(\mathrm{gin}_{\mathrm{DegRevLex}}(I))$ in any characteristic.

Thus if $\mathrm{reg}(I) = 2$, then also $\mathrm{gin}_{\mathrm{DegRevLex}}(I)$ has a 2-linear resolution, in particular $\mathrm{gin}_{\mathrm{DegRevLex}}(I)$ is generated in degree 2. □

Example 26 Consider the Borel algebra "B(t[2]^3*t[3])" We have seen in Example 10 that it is G-quadratic (in the given coordinates). We see that the presentation ideal has a 2-linear resolution:
(recall the CoCoA definition of the function "B(v)" from Example 5).

```
/**/ use S ::= QQ[t[1..3]];
/**/ I := B(t[2]^3*t[3]);
/**/ PrintRes(I);
0 --> R(-7)^6 --> R(-6)^35 --> R(-5)^84
                    --> R(-4)^105 --> R(-3)^70 --> R(-2)^21
/**/ PrintBettiDiagram(I);
          0     1     2     3     4     5
    ------------------------------------------
    2:   21    70   105    84    35     6
    ------------------------------------------
    Tot: 21    70   105    84    35     6
```

We verify that also gin(I) has a 2-linear resolution:

```
/**/ gin(I);
ideal(x[1]^2, x[1]*x[2], x[2]^2, x[1]*x[3], x[2]*x[3], x[3]^2,
x[1]*x[4], x[2]*x[4], x[3]*x[4], x[4]^2, x[1]*x[5], x[2]*x[5],
x[3]*x[5], x[4]*x[5], x[5]^2, x[1]*x[6], x[2]*x[6], x[3]*x[6],
x[4]*x[6], x[5]*x[6], x[6]^2)
/**/ PrintBettiDiagram(gin(I));
          0     1     2     3     4     5
    ------------------------------------------
    2:   21    70   105    84    35     6
    ------------------------------------------
    Tot: 21    70   105    84    35     6
```

The viceversa is not true: there are ideals defining a G-quadratic algebra and whose resolution is not 2-linear.

Example 27 We recall from Example 19 two algebras I_{22} and I_{72} introduced by Roos. We have shown that they are G-quadratic, but their corresponding $\mathrm{gin}_{\text{DegRevLex}}$ are not generated in degree 2. Therefore their resolutions, and those of their gin are not 2-linear (Proposition 25):

```
/**/ Use T ::= QQ[x,y,z,u];

/**/ I22 := ***Ideal(xz, y^2, z^2, yu + zu)***;
/**/ PrintBettiDiagram(I22);
         0    1    2    3
------------------------------
   2:    4    2    -    -
   3:    -    4    4    1
------------------------------
 Tot:    4    6    4    1

/**/ gin(I22);
ideal(x^2, x*y, y^2, x*z, y*z^2, y*z*u)
/**/ PrintBettiDiagram(gin(I22));
         0    1    2    3
------------------------------
   2:    4    4    1    -
   3:    2    5    4    1
------------------------------
 Tot:    6    9    5    1

/**/ I72 := ***Ideal(xz, y^2, yu, z^2, zu, u^2)***;
/**/ PrintBettiDiagram(I72);
         0    1    2    3
------------------------------
   2:    6    7    2    -
   3:    -    2    3    1
------------------------------
 Tot:    6    9    5    1

/**/ gin(I72);
ideal(x^2, x*y, y^2, x*z, y*z, z^2, x*u^2)
/**/ PrintBettiDiagram(gin(I72));
         0    1    2    3
------------------------------
   2:    6    8    3    -
   3:    1    3    3    1
------------------------------
 Tot:    7   11    6    1
```

5   Koszul Algebras and Hilbert Series

Now we consider a more general case: R a standard graded K-algebra (with n generators of degree 1). Therefore we can present it as $R = S/I$ with S a polynomial ring in n variables, and I a homogeneous ideal in S. Every R-module M has a minimal free resolution over R which, in general, is not finite

$$\cdots \to \bigoplus_j R(-j)^{\beta_{ij}^R(M)} \to \cdots \to \bigoplus_j R(-j)^{\beta_{0j}^R(M)} \to M \to 0.$$

We denote by $\mathcal{M}$ the unique graded maximal ideal of R. We consider $K = R/\mathcal{M}$ as an R-module and its minimal graded resolution.

Definition 28 The algebra R is said to be **Koszul** if the residue class field K has a linear R-resolution, i.e. if $\beta_{ij}^R(K) = 0$ for every $i \neq j$.

Example 29 Let $R = K[x]/(x^s)$ with $s > 1$. The minimal free resolution of K over R is

$$\cdots \to R(-2s) \to R(-s-1) \to R(-s) \to R(-1) \to K \to 0$$

where the maps are alternatively the multiplication by x and by x^{s-1}. Hence if $s = 2$ then K has a linear R-resolution, therefore R is a Koszul algebra

Since the resolution of K over R is infinite, it is impossible to show that R is Koszul with a direct computation.

Moreover it can be also hard to show that R is not Koszul, since the linearity can fail in an arbitrary high homological position.

Example 30 In [98] Roos introduces the family of quadratic non-Koszul algebras, $R_u = \mathbb{Q}[a,b,c,d,e,f]/I_u$, depending on an integer $u > 2$, where

$$I_u = (a^2,\ ab,\ b^2,\ bc,\ c^2,\ cd,\ d^2,\ de,\ e^2,\ ef,\ f^2,\ ac+ucf-df,\ ad+cf+(u-2)df)$$

All Hilbert series of R_u are $1 + 6z + 8z^2$ (independently of u). Moreover the first non-linear syzygy in the resolution of K over R_u is in homological position $u + 1$. For $u = 6$ the resolution of $\mathbb{Q}$ over R_u has Betti Diagram

0 :	1	6	28	120	496	2016	8128	32640
1 :	–	–	–	–	–	–	–	1
tot :	1	6	28	120	496	2016	8128	32641

As it is so hard to tackle Koszulness directly, we are now going to investigate for sufficient and/or necessary conditions.

Remark 31 First of all it is easy to see that

$$
\beta^R_{2j}(K) = \begin{cases} \beta^S_{1j}(R) & \text{if } j \neq 2 \\[2mm] \beta^S_{12}(R) + \binom{n}{2} & \text{if } j = 2 \end{cases}
$$

and hence the resolution of K, as an R-module, is linear up to homological position 2 if and only if R is quadratic. In particular, if R is Koszul, then R is quadratic.

Remark 32 If I is generated by monomials of degree 2 with respect to some coordinate system, then a filtration argument shows that R is Koszul. More precisely, for every subset Y of variables $R/(Y)$ has an R-linear resolution.

Moreover If I is generated by a regular sequence of quadrics, then R is Koszul (see [34, Remark 1.11,1.12] by Conca).

But not all quadratic algebras are Koszul (see Example 37 below).

We now recall a very well know tool in Commutative Algebra and see its relations with Koszul algebras.

Definition 33 Let $M = \oplus_{i\in\mathbb{Z}} M_i$ be a finitely generated graded S-module. We denote by $H_M(t)$ the **Hilbert series** of M

$$
H_M(t) = \sum_{i\in\mathbb{Z}} \dim_K(M_i) z^i \in \mathbb{Q}[|z|][z^{-1}]
$$

The Hilbert series is a rational function: more precisely $H_M(t) = \frac{h(t)}{(1-t)^d}$ where $h(t) \in \mathbb{Z}[t]$ and $0 \leq d \leq n$.

If $h(1) \neq 0$ and $h(t) = h_0 + h_1 t + \cdots + h_s t^s$ (with $h_s \neq 0$) then $h(t)$ is called **h-polynomial** of M and $(h_0, \ldots, h_s)$ is called **h-vector** of M.

Example 34 The modules we consider here are of the form S/I.

```
/**/ use S ::= QQ[x,y,z];
/**/ M := S/ideal(x^2*y +z^3, y^2 -x*z);
/**/ HilbertSeries(M);
(1 + 2*t + 2*t^2 + t^3) / (1-t)
/**/ HVector(M);
[1, 2, 2, 1]
```

To deal in a compact way with the information we get from the R-resolution of M we consider:

Definition 35 The **Poincaré series** of an R-module M is defined as

$$
P^R_M(t) = \sum_{i\in\mathbb{N}} \beta^R_i(M) z^i \in \mathbb{Q}[|z|]
$$

where $\beta^R_i(M) = \sum_j \beta^R_{ij}(M)$.

In general the Poincaré series $P_K^R(t)$ of K may be not rational, see [5] by Anick. However one has:

Proposition 36 *If R is a Koszul algebra then the Poincaré series $P_K^R(t)$ is rational. More precisely*

$$P_K^R(t) \cdot H_R(-t) = 1.$$

Indeed this equality is equivalent to R being Koszul (see for instance [57] by Fröberg).

From this fact it follows that, for a Koszul algebra R, all the coefficients in the series $1/H_R(-t)$ must be positive: this is a necessary condition that can be checked by computations (for a finite number of steps).

Example 37 Consider the ideal $I_1 = (x^2, y^2, z^2, w^2, xy + zw)$. Then $R = S/I_1$ is quadratic. Is it a Koszul algebra?

```
/**/ Use S ::= QQ[x,y,z,w];
/**/ I1 := Ideal(x^2, y^2, z^2, w^2, x*y +z*w);
/**/ H := HilbertSeries(S/I1);    H;
(1 + 4*t + 5*t^2) / (1-t)^0
/**/ HV := HVector(S/I1);    HV;
[1, 4, 5]
```

Now we check whether the first d coefficients of $1/H(-t)$ are positive (necessary condition for S/I_1 to be Koszul).

Note: we show here how to compute this for any Hilbert Series $H(t) = \frac{N(t)}{D(t)}$ even though this example has a very simple denominator $D(t) = (1 - t)^0$.

Recall that we can write $H(t)$ with $N(t)$ and $D(t) = (1 - t)^d$ coprime, and from $N(t) = 1 - t \cdot F(t)$ we have $\frac{1}{N(t)} = 1 + (t \cdot F(t)) + (t \cdot F(t))^2 + (t \cdot F(t))^3 + \dots$.

Then $\frac{1}{H(-t)} = \frac{D(-t)}{N(-t)} = D(-t) \cdot (1 + (t \cdot F(-t)) + (t \cdot F(-t))^2 + (t \cdot F(-t))^3 + \dots)$

```
/**/ QQt := RingQQt(1);    Use QQt;
/**/ D := (1-t)^(dim(S/I1));    D;
1
/**/ N := sum([HV[i]*t^(i-1) | i in 1..len(HV)]);    N;
5*t^2 +4*t +1

/**/ DD := (1-(-t))^(dim(S/I1));    DD;
1
/**/ NN := sum([HV[i]*(-t)^(i-1) | i in 1..len(HV)]);    NN;
5*t^2 -4*t +1
/**/ tF := 1 - NN;
-5*t^2 +4*t
```

and now we calculate the first $d = 10$ terms of the series $\frac{D(-t)}{N(-t)}$:

```
/**/ d := 10;
/**/ Pd := NR(DD * sum([ tF^i | i in 0..d]), [t^(d+1)]);    Pd;
-6469*t^10 -3116*t^9 -1199*t^8 -336*t^7 -29*t^6 +44*t^5 +41*t^4
+24*t^3 +11*t^2 +4*t +1
```

To make this checking more automatic we can ask CoCoA to find for us the smallest coefficient, or the lowest degree with a negative coefficient:

```
/**/ min([ CoeffOfTerm(Pd,t^i) | i in 0..d]); -- min coeff
-6469
/**/ min([ i in 0..d | CoeffOfTerm(Pd,t^i)<0]);
6
```

Then the quadratic algebra S/I_1 is *not a Koszul algebra*.

As a corollary by [96, Proposition 4.13] (Reiner and Welker) one has another necessary condition which can be useful to show that an algebra is not Koszul:

Proposition 38 *If R is a Koszul algebra then its h-polynomial h(t) has at least one real root.*

Example 39 From the example above, $I_1 = (x^2, y^2, z^2, w^2, xy + zw)$, we can also check the real roots of its h-polynomial, $1 + 4t + 5t^2$:

```
/**/ RealRoots(N);
[]
```

No real roots, therefore (we confirm) it is not Koszul.

Let's consider another example.

Example 40

```
/**/ Use S ::= QQ[a,b,c,d];
/**/ I2 := *** Ideal(ac, ad, ab - bd, a^2 + bc, b^2) ***;
/**/ HilbertSeries(S/I2);
(1 + 2*t - 2*t^2 - 2*t^3 + 2*t^4) / (1-t)^2
/**/ HV := HVector(S/I2);  HV;
[1, 2, -2, -2, 2]

/**/ use QQt;
/**/ D := (1-t)^(dim(S/I2));
/**/ N := sum([HV[i]*t^(i-1) | i in 1..len(HV)]);  N;
2*t^4 -2*t^3 -2*t^2 +2*t +1
/**/ indent(factor(N));
record[
  RemainingFactor := 1,
  factors := [2*t^4 -2*t^3 -2*t^2 +2*t +1],
  multiplicities := [1]
]
```

This means that N is irreducible over $\mathbb{Q}$, in particular it has no *rational* roots. Check for *real* roots:

```
/**/ indent(RealRoots(N), 2);
[
  Record[
    CoeffList := [-2, 2, 2, -2, -1],
    inf := -1,
    sup := -1/2
  ],
```

```
Record[
  CoeffList := [2, -2, -2, 2, 1],
  inf := -1/2,
  sup := 0
  ]
]
```

CoCoA Remark *The function "*`RealRoots`*" computes the real roots of a univariate polynomial effectively by determining isolating intervals for the real roots (and also the coefficients of polynomials with the same root in those intervals). This is indeed the hardest step in finding the real roots.*

The process of refining the interval of a real root is very fast in CoCoA ([1], Abbott), so we can ask for a higher precision (even much much higher!)

```
/**/ RRA := RealRootsApprox(N);  RRA; --> rational approx
[-852299/1048576, -112403/262144]
/**/ [DecimalStr(r) | r in RRA]; --> decimal representation
["-0.813", "-0.429"]

/**/ RRA := RealRootsApprox(N, 10^(-100)); -- max separation
/**/ [DecimalStr(r, 100) | r in RRA];


["-0.8128147866565954703146112751343851933277675752198950916305570
95138996541142118109495009626207601D369",
+"-0.4287815744562657471003815898597939890067581215721682711798170
8710644784705172132127772473866651999958"]
```

however computed, it does have a real root: so it **may** be Koszul.

Check now whether the Poincaré series has non-negative coefficients:

```
/**/ DD := Subst(D, t, -t);  tF := 1-Subst(N, t, -t);
/**/ d := 150;  --  first d coeffs in Poincare' Series
/**/ Pd := NR(DD*Sum([ tF^i | i in 0..d]), [t^(d+1)]);
/**/ min([ CoeffOfTerm(Pd,t^i) | i in 0..d]);
1
```

Then again S/I_2 **may** be Koszul. We will investigate this example again in the next sections (Example 49, Example 54).

Exercise 41 Look for Hilbert series that are not series of a Koszul algebra.

Exercise 42 Look for such Hilbert series, with the further property that the h-polynomial has a real root (harder!)

6   G-Quadratic and LG-Quadratic Algebras

In the previous sections we have seen some ways to show, by computation, that an algebra is *not Koszul*. We introduce now some computable conditions which are stronger than being Koszul.

Proposition 43 *If R is G-quadratic (Definition 9), then R is a Koszul algebra.*

This follows from Remark 32 and from the standard deformation argument showing that $\beta_{ij}^R(K) \leq \beta_{ij}^A(K)$ with $A = S/\operatorname{in}_\tau(I)$. For details see, for instance, [28, Sect. 3] (Bruns and Conca).

Example 44 Consider the polynomial ring $K[t_1, t_2, t_3]$ over a field K, and let R be the K-subalgebra of $K[t_1, t_2, t_3]$ generated by the *general lex-segment set* (introduced in Sect. 2.1) $L(t_1^2, t_2 t_3) = \{t_1^2, t_1 t_2, t_1 t_3, t_2^2, t_2 t_3\}$. As seen in Sect. 2 we calculate the presentation ideal of R:

CoCoA Remark *The function "LexSegmentIdeal(L)" returns the smallest lex-segment ideal (which starts from t_1^d) containing all the power-products in the given list L.*

```
/**/ use QQ[t[1..3]];
/**/ L := gens(LexSegmentIdeal([t[2]*t[3]]));    L;
[t[2]*t[3], t[2]^2, t[1]*t[3], t[1]*t[2], t[1]^2]

/**/ R ::= QQ[x[1..len(L)]];
/**/ M := transposed(mat([ exponents(f) | f in L ]));
/**/ I := toric(R, M);    gens(I);
[-x[3]*x[4]+x[1]*x[5], -x[4]^2+x[2]*x[5], x[2]*x[3]-x[1]*x[4]]
/**/ GBasis(I);
[-x[4]^2+x[2]*x[5], -x[3]*x[4]+x[1]*x[5], x[2]*x[3]-x[1]*x[4]]
```

Then R is G-quadratic, thus Koszul.

We now introduce the weaker concept of LG-quadratic algebra (where the "L" stands for "Lifting").

Definition 45 A K-algebra R is **LG-quadratic** if there exist a G-quadratic algebra A and a regular sequence of linear forms $y_1, \ldots, y_c$ such that

$$R \simeq A/(y_1, \ldots, y_c)$$

Remark 46 The notion of being LG-quadratic is clearly weaker than being G-quadratic, but still implies Koszulness (see for example [37, Sect. 3.2] by Conca et al.).

The following is an example of LG-quadratic algebra.

Example 47 Let

$$R = K[a, b, c, d]/(a^2 - bc, \ d^2, \ cd, \ b^2, \ ac, \ ab).$$

We first check whether R is G-quadratic. We look for a quadratic Gröbner basis (naive approach):

```
/**/ use S ::= QQ[a,b,c,d]; -- default order: DegRevLex
/**/ I := *** Ideal(a^2-bc, d^2, cd, b^2, ac, ab) ***;
```

```
/**/ GBasis(I);
[d^2, c*d, a*c, b^2, a*b, a^2 -b*c, b*c^2]

----> try another order:
/**/ SDL ::= QQ[a,b,c,d], DegLex;
/**/ GBasis(ideal(BringIn(SDL, gens(I))));
[d^2, c*d, b^2, a*c, a*b, a^2 -b*c, b*c^2]
```

CoCoA Remark *The function "*`BringIn(f)`*" constructs, if possible, an algebra homomorphism from the ring of the polynomial f into the current ring, maintaining the names of the indeterminates occurring in f (for example a $\mapsto$ a) independently of their respective index in the domain and codomain. It can also be applied, as here, to a list of polynomials. It isn't very efficient, but it's quite convenient!*

We haven't found a Gröbner basis made of quadrics, and we can indeed prove that R is not G-quadratic! We now show that there is no quadratic monomial ideal with the same h-vector as I, therefore no Gröbner basis of I may be made of quadrics in any coordinates. We follow a "brute force" strategy, i.e. we try all the ideals generated by 6 quadratic monomials in S:

```
/**/ HV := HVector(S/I);   HV;
[1, 3, 0, -3]
/**/ Mons := gens(ideal(a,b,c,d)^2); -- all quadratic monomials
/**/ G := subsets(Mons,6); -- all subsets of M of cardinality 6
/**/ foreach L in G do
       if HVector(S/ideal(L)) = HV then
         print "Found! ", L;
         break;
       endif;
     endforeach;
--> no monomial ideal is found
```

But maybe R is LG-quadratic. So we try looking for a G-quadratic algebra $A = S_5/J$, with $S_5 = \mathbb{Q}[a, b, c, d, e]$ (one more indeterminate), and a linear non-zerodivisor l such that $R = A/(l)$.

First we look for a quadratic monomial ideal, the candidate $in_\sigma(J)$, with h-vector $[1, 3, 0, -3]$: (again by "brute force" search)

```
/**/ use S5 ::= QQ[a,b,c,d,e];
/**/ Mons := gens(ideal(a,b,c,d,e)^2);
/**/ G := subsets(Mons, 6);

/**/ foreach L in G do
       if HVector(S5/ideal(L)) = HV then
         print "Found! ", L;
         break;
       endif;
     endforeach;
```

```
Found! [e^2, d*e, c*d, b*c, a*b, a^2]
```

So S/I might be LG-quadratic. We modify the loop above so to create the list "ALL" containing the generators of *all* monomial ideals with h-vector "HV".

```
/**/ ALL := [];
/**/ foreach g in G do
       if HVector(S5/ideal(g)) = HV then
         append(ref ALL, g);   -- ref  indicates ALL is modified
       endif;
     endforeach;
/**/ len(ALL);
60
```

there are 120 permutations of the indeterminates a, b, c, d, e. How many truly different ideals have we found? Take one and see how many permutations of indeterminates fix it. (we write a function applying a permutation "Perm" to an ideal J)

```
/**/ C := ideal(ALL[1]);   C;  -- one of the candidates
ideal(e^2, d*e, c*d, b*c, a*b, a^2)

/**/ define map(Perm, J) -- returns the image of J under Perm
       phi := PolyAlgebraHom(RingOf(J), RingOf(J), Perm);
       return ideal(apply(phi, gens(J)));
     enddefine;

/**/ [ perm in permutations([a,b,c,d,e]) | map(perm, C) = C];
[[a, b, c, d, e], [e, d, c, b, a]]
```

which means that the permutations that fix C are the identity and $(a, e)(b, d)$. So there are $120/2 = 60$ elements in the orbits of C, which implies that there is essentially a unique monomial ideal in 5 indeterminates with the same h-vector as I. In other words: for all G-quadratic algebras S_5/J with h-vector $[1, 3, 0, -3]$, the ideal J has the same initial ideal up to permutations of the indeterminates.

Going back to our search, we need to find a G-quadratic $A = S_5/J$ and a linear non-zerodivisor l such that $R \simeq A/(l)$. This might be a quite a tricky task: the idea is that we modify the generators slightly so that the Gröbner basis has the same initial monomials, but enough to allow l to be a non-zerodivisor.

NOTE: this is *hard* work, even when you know you *can* find it (which is usually not the case while you're doing research...). We show here the "final step" of the search. We add to some generators in I a (small) multiple of the new indeterminate e (playing the role of the linear non-zerodivisor l), and see if/when we get h-vector $[1, 3, 0, -3]$ and a GBasis of quadrics.

```
/**/ use S5_ebacd ::= QQ[e,b,a,c,d];

/**/ I := *** Ideal(a^2-bc, d^2, cd, b^2, ac, ab) ***;
/**/ G := gens(I);

/**/ foreach T in tuples(1..6, 3) do
       GG := G;
```

```
        GG[T[1]] := GG[T[1]]+e*a;
        GG[T[2]] := GG[T[2]]+e*b;
        GG[T[3]] := GG[T[3]]+e*b;
        if  HVector(RingOf(I)/ideal(GG)) = [1,3,0,-3] then
          print max([ deg(F) | F In GBasis(ideal(GG))]), " ";
          println GG;
        endif;
      endforeach;
3 [2*e*b +e*a +a^2 -b*c, d^2, c*d, b^2, a*c, b*a]
6 [2*e*b +a^2 -b*c, e*a +d^2, c*d, b^2, a*c, b*a]
2 [e*b +a^2 -b*c, d^2, c*d, e*b +b^2, a*c, e*a +b*a]
2 [e*b +a^2 -b*c, d^2, c*d, e*b +b^2, a*c, e*a +b*a]
```

so we have found the desired ideal J:

```
/**/ J := ideal(e*b +a^2 -b*c, d^2, c*d,
                             e*b +b^2, a*c, e*a +b*a);
/**/ GBasis(J);
[d^2, c*d, a*c, e*a +b*a, e*b +a^2 -b*c, b^2 -a^2 +b*c]
/**/ HVector(RingOf(J)/J);
[1, 3, 0, -3]
/**/ J : ideal(e) = J;
true
```

In conclusion: $A = S_5/J$ is G-quadratic, and e is a non-zerodivisor for A. This shows that $R \simeq A/(e)$ is LG-quadratic.

Remark 48 Summing up we have the following implications:

$$\text{G-quadratic} \overset{1}{\Longrightarrow} \text{LG-quadratic} \overset{2}{\Longrightarrow} \text{Koszul} \overset{3}{\Longrightarrow} \text{quadratic} \tag{1}$$

In Example 37 we showed that $R = S/I_1$ is a quadratic algebra, but not Koszul. Thus implication 3 is strict.

In Example 47 we showed that $R = K[a, b, c, d]/(a^2 - bc, d^2, cd, b^2, ac, ab)$ is LG-quadratic, but not G-quadratic. Therefore implication 1 is strict.

In the following example we give an algebra which is not LG-quadratic, and we will see in Example 54 that this algebra is Koszul, thus proving that also implication 2 is strict.

Example 49 Let $R = K[a, b, c, d]/(ac, ad, ab - bd, a^2 + bc, b^2)$. The h-polynomial does not change under lifting with regular sequences of linear forms, therefore in order to claim that R is not LG-quadratic it is enough to show that there is no algebra with quadratic monomial relations with the same h-polynomial as R. First we check that it is not quadratic, as we did in Example 47

```
/**/ use S ::= QQ[a,b,c,d];
/**/ I2 := ideal(a*c, a*d, a*b -b*d, a^2 +b*c, b^2);
/**/ HV2 := HVector(S/I2);   HV2;
[1, 2, -2, -2, 2]

/**/ Mons := gens(ideal(indets(S))^2);
```

```
/**/ G := subsets(Mons, 5);  len(G); -- binomial(10, 5);
252

/**/ foreach g in G do
       if HVector(S/ideal(g)) = HV2 then
         print " !! FOUND !! ";
       else print "0";
       endif;
     endforeach;
0000000000000000000000000000000000000000000000000...(252 times)
```

so $R = S/I_2$ is NOT G-quadratic.

To check that it is not LG-quadratic we show that there are no monomial ideals J (therefore no quadratic Gröbner basis, and no G-quadratic algebras) whose h-vector is $(1, 2, -2, -2, 2)$. The ideal J must have codimension 2 and 5 generators, thus the ring has at most 7 indeterminates, call them $a, b, \ldots, g$. Moreover such J, having codimension 2, is contained in, say, (a, b), thus its 5 generators are a subset of the generators of the ideal $(a, b)(a, b, c, d, e, f, g)$. (an easier search than among all $\binom{8}{2}$ quadratic monomials). We look for it.

```
/**/ Use S7 ::=   QQ[a,b,c,d,e,f,g];
/**/ Mons := gens(ideal(a,b) * ideal(a,b,c,d,e,f,g));
/**/ G := subsets(Mons, 5);
/**/ len(G);  -- binomial(13, 5);
1287

/**/ foreach g in G Do
       if HVector(S7/Ideal(g)) = HV2 then
         print "Found";
       endif;
     endforeach;
```

it prints nothing, so S/I_2 is NOT LG-quadratic.

Exercise 50 Look for monomial ideals with the same h-vector as the algebra of Example 47, with 6–9 indets (more than 9 is not possible!) You should find two new ideals, one in 6 and one in 7 indeterminates (up to permutations).

Exercise 51 In Example 49 we proved that the algebra R is not LG-quadratic by showing that there is no quadratic monomial ideal with the same h-vector as R.

OPEN PROBLEM: find a non-LG-quadratic Koszul algebra with an admissible h-vector.

7   Koszul Filtrations

We introduce in this section a powerful tool to prove that an algebra is Koszul. The idea, due to Conca et al. [36], is to construct a special filtration of ideals of R.

Definition 52 Let R be a standard graded K-algebra. A **Koszul filtration** of R is a set F of ideals of R such that:

1. Every ideal in F is generated by elements of degree 1.
2. The zero ideal (0) and the maximal ideal $\mathcal{M}$ are in F.
3. For every non-zero ideal $I \in F$, there exists $J \in F$ such that $J \subset I$, I/J is cyclic, and $J : I \in F$.

The existence of a Koszul filtration of R implies the Koszulness of R, but even more (see [36, Proposition 1.2])

Lemma 53 *Let F be a Koszul filtration of R. Then all the ideals in F have a linear resolution. In particular R is Koszul.*

Now we can prove that the non-LG-quadratic algebra of Example 49 is Koszul.

Example 54 Let $R = K[a,b,c,d]/(ac, ad, ab - bd, a^2 + bc, b^2)$. We have seen that R is neither G-quadratic, nor LG-quadratic. We prove that R is Koszul by showing that there exists a Koszul Filtration of F.

CoCoA Remark *we can work directly in a quotient ring R. Its elements are printed within "(. .)" to highlight they are classes. The special functions for polynomial rings, namely "*indets*", "*deg*",..., do not work for R and its elements.*

```
/**/ use S ::= QQ[a,b,c,d];
/**/ I2 := ideal(a*c, a*d, a*b -b*d, a^2 +b*c, b^2);
/**/ R := S/I2;
/**/ use R;   -- now we work directly in S/I2

/**/ M := ideal(a,b,c,d); -- ideal in R
/**/ J1 := ideal(a,c,d);
/**/ J1:M;
ideal((d), (c), (b), (a))

/**/ J2 := ideal(c,d);
/**/ J2:J1;
ideal((d), (c), (b), (a))

/**/ J3 := ideal(c);
/**/ J3:J2;
ideal((c), (a), (b*d))
/**/ ideal((c), (a), (b*d)) = ideal((c), (a));
true
```

...and so on. Now we write a function which verifies whether a filtration, and in particular

$$F = \{(a,b,c,d), (a,c,d), (c,d), (a,c), (c), (a), (0)\},$$

is a Koszul Filtration for R.

Following the construction, we make the relations explicit representing the filtration F in CoCoA as the set of the triples $(I, J, J : x)$ where $J + x = I$

```
/**/ F := [-- I (= J+(x))      J          J:x (= J:I)
          [[a, b, c, d],  [a, c, d],  [a, b, c, d]],
          [[a, c, d],     [c, d],     [a, b, c, d]],
          [[c, d],        [c],        [a, c]],
          [[a, c],        [c],        [a, c, d]],
          [[c],           [],         [a]],
          [[a],           [],         [c, d]]
         ];
```

first we write a function checking that any triple in such a list satisfies the conditions $(J + (x), \ J, \ J : x)$.

```
define IsCorrectTriple(R, T)   --  T = [I, J, J:x]
  I := T[1];    J := T[2];    Jx := T[3];
  x := diff(I, J);
  if    len(x)=1
    and IsSubset(J, I)
    and ideal(R,J) : ideal(x) = ideal(R,Jx)   then
    return true;
  endif;
  return false;
enddefine; -- IsCorrectTriple

/**/ [ IsCorrectTriple(R, T) | T in F];
[true, true, true, true, true, true]
```

CoCoA Remark *When calling the function "`ideal(L)`", constructing an ideal I from a list of generators L, CoCoA stores in I the ring R it belongs to, that is the ring of all the elements in L. In case L might be empty, we need to pass this information using the unequivocal syntax "`ideal(R, L)`".*

Now we can make a handy function checking whether the list "F" represents a filtration:

```
/**/ define IsKoszulFiltration(R, F)
        foreach T in F do
          if not(IsCorrectTriple(R, T)) then
            return false;
          endif;
        endforeach;
        F1 := [ T[1] | T in F ];
        F2 := [ T[2] | T in F and T[2]<>[] ];
        F3 := [ T[3] | T in F ];
        if not(IsSubset(F2, F1)) then return false; endif;
        if not(IsSubset(F3, F1)) then return false; endif;
        return true;
      enddefine; -- IsKoszulFiltration

/**/ IsKoszulFiltration(CurrentRing, F);
true
```

It can be useful to write filtrations containing more ideals, since in this way one gets more ideals with a linear resolution (Lemma 53).

Example 55 We write another filtration of the algebra $R = S/I_2$ considered above, in Example 54.

```
/**/ F := [
        [[a, b, c, d],   [b, c, d],   [a, b, c, d]],
        [[a, b, c, d],   [a, c, d],   [a, b, c, d]],
        [[a, b, c, d],   [a, b, d],   [a, b, d]],
        [[a, b, c, d],   [a, b, c],   [a, b, c]],
        [[a, b, d],   [a, d],   [a, b, c, d]],
        [[a, b, d],   [a, b],   [a, b]],
        [[b, c, d],   [c, d],   [a, b, c, d]],
        [[a, c, d],   [c, d],   [a, b, c, d]],
        [[a, c, d],   [a, d],   [a, b, d]],
        [[a, c, d],   [a, c],   [a, b, c]],
        [[a, b, c],   [a, c],   [a, b, c, d]],
        [[a, b, c],   [a, b],   [a, b]],
        [[a, d],   [d],   [b, c, d]],
        [[a, d],   [a],   [a, b]],
        [[c, d],   [d],   [a, d]],
        [[c, d],   [c],   [a, c]],
        [[a, c],   [c],   [a, c, d]],
        [[a, c],   [a],   [a, b]],
        [[a, b],   [a],   [a, b, c, d]],
        [[d],   [ ],   [a]],
        [[a],   [ ],   [c, d]],
        [[c],   [ ],   [a]]
        ];
```

```
/**/ IsKoszulFiltration(CurrentRing, F);
true
```

By definition a Koszul filtration must contain a complete flag of ideals

$$I_0 = (0) \subset I_1 \subset \cdots \subset I_{n-1} \subset I_n = M$$

such that every I_i is minimally generated by i linear forms. Consider the special case in which F consists of just a single flag:

Definition 56 A **Gröbner flag** of R is a Koszul filtration that consists of a single complete flag of ideals of linear forms. More explicitly, a Gröbner flag is a filtration of the form $F = \{I_0, I_1, \ldots, I_n\}$ where

$$I_0 = (0) \subset I_1 \subset \cdots \subset I_{n-1} \subset I_n = M \quad \text{and} \quad I_{i-1} : I_i \in F, \ i = 1, .., n.$$

The following proposition shows the importance of the existence of a Gröbner flag:

Proposition 57 *If R has a Gröbner flag, then R is G-quadratic ([35, Theorem 2.4] by Conca et al.).*

Remark 58 Various families of Algebras with Koszul filtration or Gröbner flag are described in [35] and [36].

In particular we recall that a ring of coordinates of a set of $\leq 2n$ points in general linear position in $\mathbb{P}^n$ has a Gröbner flag, and any general Gorenstein Artinian algebra with socle in degree 3 has a Koszul filtration.

On the other hand the same paper also shows that there are Koszul algebras without Koszul filtrations, and G-quadratic algebras without Gröbner flags (see [35, pp. 100 and 101]).

Exercise 59 Take a binomial quadratic ideal I in a polynomial ring P in 4 or 5 indeterminates, and look for a Koszul Filtration for P/I. If one exists, then P/I is a Koszul algebra.

Exercise 60 (Difficult) Write a function which (tries to) construct a filtration. [Hint: start with the maximal ideal . . . .]

8   Complete Intersection of Quadrics

In this section we study the algebras defined by ideals generated by complete intersections of homogeneous polynomials of degree 2, and show how they fit in our Koszul algebras scenario.

Proposition 61 *Let $\{q_1, \ldots, q_r\}$ be a complete intersection of forms of degree 2 in $K[x_1, \ldots, x_n]$. Then the algebra $R = K[x_1, \ldots, x_n]/(q_1, \ldots, q_r)$ is LG-quadratic.*

Proof Let $T = K[w_1, \ldots, w_r, x_1, \ldots, x_n]$ and let $J = (w_1^2 + q_1, \ldots, w_r^2 + q_r)$. Fix on T the lexicographic term order: this satisfies $w_i > x_j$ for every i, j. Then the generators of J, whose leading terms are coprime, are a Gröbner basis, and therefore $A = T/J$ is G-quadratic. Now note that $w_1, \ldots, w_r$ is a regular sequence in A because

$$A/(w_1, \ldots, w_r) \simeq R \quad \text{and} \quad \dim(A) - \dim(R) = r$$

therefore R is LG-quadratic.   □

Example 62 We show that

$$R = S/I = \mathbb{Q}[x, y, z]/(x^2 + yz, \ y^2 + xz, \ z^2 + xy)$$

is defined by a complete intersection of quadrics.

```
/**/ use S ::= QQ[x,y,z];
/**/ L := [x^2 + y*z,   y^2 + x*z,   z^2 + x*y];
/**/ I := ideal(L);
/**/ dim(S/I);
0
```

then I is a complete intersection (in particular R is artinian), and therefore, by Proposition 61, R is LG-quadratic.

We show it by the explicit computation described in the proof.

```
/**/ use T ::= QQ[w[1..3], x,y,z], lex;
/**/ L_T := BringIn(L); -- shortcut for "PolyAlgebraHom"
/**/ J := ideal([w[i]^2 + L_T[i] | i in 1..3]); J;
ideal(w[1]^2 +x^2 +y*z, w[2]^2 +x*z +y^2, w[3]^2 +x*y +z^2)

/**/ dim(T/J); -- w[1],w[2],w[3] regular sequence
3

/**/ GBasis(J);
[w[3]^2 +x*y +z^2, w[2]^2 +x*z +y^2, w[1]^2 +x^2 +y*z]
```

Thus T/J is G-quadratic, and that implies that $R \simeq (T/J)/(w_1, w_2, w_3)$ is LG-quadratic.

Is a complete intersection generated by quadrics also G-quadratic? The answer is "no" and we are going to prove that the algebra of Example 62 is a counter-example. So this again shows that the implication "G-quadratic $\Longrightarrow$ LG-quadratic" is strict (the first example was 47).

We recall a result for G-quadratic artinian algebras and its simple proof.

Proposition 63 *Let S/I be a G-quadratic artinian algebra. Then I contains the square of a linear form.*

Proof By definition, there is some change of coordinates g and some term order σ such that the σ-Gröbner basis G of $g(I)$ is made of quadrics. Being artinian $(\dim(S/I) = 0)$ for each indeterminate x_i there must be a polynomial in G whose leading term is x_i^2; in particular, for the σ-smallest indeterminate x_j, the only homogeneous polynomial with leading term x_j^2 is x_j^2 itself. Then $g^{-1}(x_j)$ is a linear form whose square is in I. □

Example 64 Consider again $R = \mathbb{Q}[x, y, z]/I$ with $I = (x^2 + yz, y^2 + xz, z^2 + xy)$, which is LG-quadratic.

We have seen that R is artinian; does I contain the square of a linear form? Consider a linear form $\ell = x + ay + bz$ and the polynomial system derived by imposing that ℓ^2 is in I: can we find solutions for a, b?

```
/**/ Use Sx ::= QQ[a,b, r,s,t, x,y,z];
/**/ q := [x^2 + y*z, y^2 + x*z, z^2 + x*y];

/**/ l := x + a*y + b*z; -- generic linear form
/**/ Eq := l^2 - (r*q[1] +s*q[2] +t*q[3]); -- l^2 in (q1,q2,q3)
/**/ CoeffPPs := CoefficientsWRT(Eq, [x,y,z]);
/**/ indent(CoeffPPs); -- all coefficients wrt x,y,z must be 0
[
  record[PP := x^2, coeff := -r +1],
  record[PP := x*y, coeff := 2*a -t],
  record[PP := y^2, coeff := a^2 -s],
  record[PP := x*z, coeff := 2*b -s],
  record[PP := y*z, coeff := 2*a*b -r],
  record[PP := z^2, coeff := b^2 -t]
]
/**/ System := [ R.coeff | R In CoeffPPs ]; System;
```

```
[-r +1, 2*a -t, a^2 -s, 2*b -s, 2*a*b -r, b^2 -t]
/**/ 1 IsIn ideal(System);
true
```

which proves that there is no solution for the system $\begin{cases} -r+1 & = 0 \\ 2a-t & = 0 \\ a^2-s & = 0 \\ 2b-s & = 0 \\ 2ab-r & = 0 \\ b^2-t & = 0 \end{cases}$ (not even in

the algebraic closure of $\mathbb{Q}$). In other words, no square of a linear form is in I, then $R = S/I$ is not G-quadratic by Proposition 63.

Now we want to see that the same holds for 3 random quadrics in $\mathbb{Q}[x, y, z]$. We can use a different approach: consider the matrix M associated to a given quadric Q and recall that Q is the square of a linear form if and only if $\text{rank}(M) = 1$, which is equivalent to saying that all the 2-minors of M must be zero.

Example 65 Consider the algebra $R = \mathbb{Q}[x, y, z]/I$ defined by the ideal generated by three generic quadrics in three indeterminates. Instead of working with the quadrics we can consider three random symmetric matrices. We start by writing a function for generating $N \times N$ random symmetric matrices over $\mathbb{Q}$:

```
define RandomSymmMat(N)
  M := mat([[random(-10,10) | i In 1..N] | j In 1..N ]);
  return M + transposed(M);
enddefine;
```

Let M_1, M_2, M_3 be the matrices associated to the three quadratic generators of I. A quadratic form Q is in I if and only if Q is a linear combination of the three quadrics. So to see if Q is the square of a linear form, it is enough to see if all the 2-minors of the matrix associated to Q are zero:

```
/**/ Use S ::= QQ[a,b,c];
/**/ M1 := RandomSymmMat(3);   M1;
matrix(QQ,
  [[-14, 2, 6],
   [2, 10, 5],
   [6, 5, 20]])
/**/ M2 := RandomSymmMat(3);   M2;
matrix(QQ,
  [[12, -5, -6],
   [-5, -18, 2],
   [-6, 2, -2]])
/**/ M3 := RandomSymmMat(3);   M3;
matrix(QQ,
  [[-18, -1, -3],
   [-1, 8, 2],
   [-3, 2, -20]])

/**/ QM := a*Mat(S,M1) + b*Mat(S,M2) + c*Mat(S,M3);   QM;
matrix( /*RingWithID(652,"QQ[a,b,c]")*/
```

```
  [[-14*a +12*b -18*c,  2*a -5*b -c,  6*a -6*b -3*c],
   [2*a -5*b -c, 10*a -18*b +8*c,  5*a +2*b +2*c],
   [6*a -6*b -3*c,  5*a +2*b +2*c, 20*a -2*b -20*c]]])
/**/ J2x2 := ideal(minors(QM,2));  indent(J2x2);


ideal(
  -144*a^2 +392*a*b -241*b^2 -288*a*c +410*b*c -145*c^2,
  -82*a^2 +74*a*b -6*b^2 -106*a*c -33*b*c -39*c^2,
  -50*a^2 +147*a*b -118*b^2 -19*a*c -18*b*c +22*c^2,
  -82*a^2 +74*a*b -6*b^2 -106*a*c -33*b*c -39*c^2,
  -316*a^2 +340*a*b -60*b^2 -44*a*c -240*b*c +351*c^2,
  10*a^2 -86*a*b +22*b^2 -57*a*c +120*b*c +26*c^2,
  -50*a^2 +147*a*b -118*b^2 -19*a*c -18*b*c +22*c^2,
  10*a^2 -86*a*b +22*b^2 -57*a*c +120*b*c +26*c^2,
  175*a^2 -400*a*b +32*b^2 -60*a*c +336*b*c -164*c^2
)
```

So we need to check if the following system admits solutions

$$\begin{cases} -144a^2 + 392ab - 241b^2 - 288ac + 410bc - 145c^2 & = 0 \\ -82a^2 + 74ab - 6b^2 - 106ac - 33bc - 39c^2 & = 0 \\ \vdots \end{cases}$$

```
/**/ GBasis(J2x2);
[a^2, -a*b, -b^2, -a*c, b*c, -c^2]
```

The Gröbner basis gives us an equivalent system, and it tells that $(0, 0, 0)$ (i.e. the null linear combination) is the only one whose square is in I (we could get the same conclusion by seeing that "dim(S/J2x2);" returns 0).

We can conclude, by Proposition 63, that R is not G-quadratic.

Now that we know how to tackle this problem we ask: what happens with more random quadrics? In the next example we consider 4 random quadrics in $\mathbb{Q}[x, y, z]$.

Example 66 Let q_1, q_2, q_3, q_4 be 4 random quadrics in $\mathbb{Q}[x, y, z]$. In this case we can find a linear form ℓ such that ℓ^2 is in the ideal generated by the quadrics.

```
/**/ Use S ::= QQ[a,b,c,d];
/**/ M := [ RandomSymmMat(3) | i in 1..4 ];  -- 4 matrices
/**/ QM := sum([ indet(S,i) * mat(S,M[i]) | i in 1..4 ]);  QM;


matrix( /*RingWithID(186,"QQ[a,b,c,d]")*/
  [[12*a -18*b -2*c +16*d, -5*a -b +6*c +5*d, -6*a -3*b -9*c +2*d
   ],
   [-5*a -b +6*c +5*d, -18*a +8*b -2*c -6*d, 2*a +2*b +20*d],
   [-6*a -3*b -9*c +2*d, 2*a +2*b +20*d, -2*a -20*b +14*c -20*d]]])


/**/ j2x2 := ideal(minors(QM,2));  indent(J2x2);
```

```
ideal(
  -241*a^2 +410*a*b -145*b^2 +72*a*c +32*b*c -32*c^2 -310*a*d
     +246*b*d -80*c*d -121*d^2,
  -6*a^2 -33*a*b -39*b^2 -13*a*c +5*b*c +54*c^2 +312*a*d -311*b*d
     -7*c*d +310*d^2,
  -118*a^2 -18*a*b +22*b^2 -162*a*c +78*b*c -18*c^2 -90*a*d -44*b
     *d +70*c*d +112*d^2,
  -6*a^2 -33*a*b -39*b^2 -13*a*c +5*b*c +54*c^2 +312*a*d -311*b*d
     -7*c*d +310*d^2,
  -60*a^2 -240*a*b +351*b^2 +64*a*c -266*b*c -109*c^2 -248*a*d
     +52*b*d +300*c*d -324*d^2,
  22*a^2 +120*a*b +26*b^2 -64*a*c -116*b*c +84*c^2 +206*a*d -24*b
     *d +130*c*d -140*d^2,
  -118*a^2 -18*a*b +22*b^2 -162*a*c +78*b*c -18*c^2 -90*a*d -44*b
     *d +70*c*d +112*d^2,
  22*a^2 +120*a*b +26*b^2 -64*a*c -116*b*c +84*c^2 +206*a*d -24*b
     *d +130*c*d -140*d^2,
  32*a^2 +336*a*b -164*b^2 -248*a*c +152*b*c -28*c^2 +292*a*d
     -120*b*d -44*c*d -280*d^2
)


/**/ dim(S/J2x2);
1
/**/ multiplicity(S/J2x2);
4
```

which means there are 4 such linear forms. We want to find them explicitly, but this
time the Gröbner basis is far less "readable":

```
/**/ indent(GBasis(J2x2));


[
  -482122639284*a^2 +165307519484*a*d -676396836248*b*d
     +96565732488*c*d +932336337887*d^2,
  482122639284*a*b -1792089671654*a*d +2149384643246*b*d
     +1382917368*c*d -1521646989413*d^2,
  482122639284*b^2 -5242064321144*a*d +5548490268764*b*d
     +213501069096*c*d -2990726347559*d^2,
  -160707546428*a*c +666298696782*a*d -677123463176*b*d
     +90413198422*c*d +231475426489*d^2,
  -160707546428*b*c +1465950864728*a*d -1507341119842*b*d
     +108626974914*c*d +589040790531*d^2,
  160707546428*c^2 -729292812056*a*d +896523161576*b*d
     +15563055452*c*d -143090171789*d^2
]
```

So we use another strategy: first we define a function that, given a symmetric matrix,
calculates the corresponding quadratic form.

```
define QForm(M, X)
  if not(IsSymmetric(M)) then error("M must be symmetric");
     endif;
  Sx := RingOf(X[1]);
```

```
    return (RowMat(X) * Mat(Sx,M) * ColMat(X)) [1,1];
  enddefine; -- QForm
```

then we write explicitly the system obtained by imposing that the square of the linear form is in the ideal generated by the 4 forms. We use the lexicographic order because the resulting Gröbner basis has a particular nice shape (see The Shape Lemma: in Kreuzer-Robbiano book [69, Theorem 3.7.25])

```
/**/ Use Sx ::= QQ[a,b,c,d,e,f, x,y,z], lex;
/**/ QF := [ QForm(Mat(Sx,M[i]), [x,y,z]) | i in 1..4];
/**/ L  := x + a*y + b*z;
/**/ Eq := L^2 - (c*QF[1] + d*QF[2] + e*QF[3] + f*QF[4]);  Eq;


a^2*y^2 +2*a*b*y*z +2*a*x*y +b^2*z^2 +2*b*x*z -12*c*x^2 +10*c*x*y
+12*c*x*z +18*c*y^2 -4*c*y*z +2*c*z^2 +18*d*x^2 +2*d*x*y +6*d*x*z
-8*d*y^2 -4*d*y*z +20*d*z^2 +2*e*x^2 -12*e*x*y +18*e*x*z +2*e*y^2
-14*e*z^2 -16*f*x^2 -10*f*x*y  -4*f*x*z +6*f*y^2 -40*f*y*z +20*f*
   z^2 +x^2


/**/ CPP := CoefficientsWRT(Eq, [x,y,z]);    indent(CPP);


[
  record[PP := x^2, coeff := -12*c +18*d +2*e -16*f +1],
  record[PP := x*y, coeff := 2*a +10*c +2*d -12*e -10*f],
  record[PP := x*z, coeff := 2*b +12*c +6*d +10*e -4*f],
  record[PP := y^2, coeff := a^2 +18*c -8*d +2*e +6*f],
  record[PP := y*z, coeff := 2*a*b -4*c -4*d -40*f],
  record[PP := z^2, coeff := b^2 +2*c +20*d -14*e +20*f]
]

/**/ System := [ R.coeff | R in CPP ];    indent(System);


[
  -12*c +18*d +2*e -16*f +1,
  2*a +10*c +2*d -12*e -10*f,
  2*b +12*c +6*d +18*e -4*f,
  a^2 +18*c -8*d +2*e +6*f,
  2*a*b -4*c -4*d -40*f,
  b^2 +2*c +20*d -14*e +20*f
]


/**/ I := ideal(System);
/**/ not(1 IsIn I); --  <==> solutions over CC?
true
/**/ GB := GBasis(I);  indent(GB);


[
  -166459332033282760350364352*c -170937253404316394726722247096*f
      ^3 +629116687902293248944127284*f^2 ...
```

```
-83229666016641380175182176*b -11442255542819315861369506056*f
    ^3 -6937521384868448839162742412*f^2 ...
-83229666016641380175182176*a +104713550384428402305433125672*f
    ^3 -30677447660017288996021160148*f^2 ...
166459332033282760350364352*d +13442410660404320513885562024*f
    ^3 -4661473973995372457540395972*f^2 ...
-166459332033282760350364352*e +18419343901049047788936575640*f
    ^3 -4206264491820757181215926684*f^2 ...
2374079062216368*f^4 -989057209805952*f^3 +149405242988472*f^2
    -6432565462160*f +40176886607
]
```

This shows the particular shape of the lex-Gröbner basis of a radical 0-dimensional ideal in x_n-normal position (i.e. whose solutions have all different n-th coordinate): one polynomial only in x_n of degree equal to the multiplicity, and polynomials of the form $x_i - g_i(x_n)$ for $i = 1, \ldots, n-1$. This shape illustrates that the system has indeed 4 solutions over complex field.

To make these solutions explicit, we first check if the univariate polynomial (in f) has rational roots (in this case not). Then we look for an *approximation* of the real roots.

```
/**/ indent(factor(last(GB)));
record[
  RemainingFactor := 1,
  factors := [2374079062216368*f^4 -989057209805952*f^3
          +149405242988472*f^2 -6432565462160*f +40176886607],
  multiplicities := [1]
]
--> irreducible over QQ
--> may be 0, 2, or 4 real roots
/**/ RRA := RealRootsApprox(last(GB));   RRA;
[31387/4194304, 951147/16777216]
/**/ [ DecimalStr(R) | R in RRA ];
["0.007", "0.057"]
/**/ RRA := RealRootsApprox(last(GB), 10^(-120));
/**/ indent([ DecimalStr(R, 120) | R In RRA ]);

[
  "0.007483230640320123143892531706334166962268878388614627292915
  632929572635202697446401252046578486294713164376415211162997",
  "0.056692738558800478390123429343687008627500825335962253017322
  120769982036641831723549814422414244461868630763925590124616"
]
```

and we can verify how good these approximate roots are:

```
/**/ [ FloatStr(subst(last(GB), f, point)) | point in RRA ];
["-4.0413*10^(-145)", "6.4691*10^(-145)"]
```

then we substitute these real solutions to f in the other polynomials in "GB" to get the values of the other coordinates.

...or, again, we can ask CoCoA to do all this for us (even the definition of the ring!)

```
/**/ Sol := RationalSolve(System); -- solutions over QQ?
/**/ Sol; --> none
[]
/**/ Sol := ApproxSolve(System); -- solutions over RR --> 2
/**/ indent([ [ DecimalStr(X) | X in point ] | point in Sol ]);
[
  ["-0.030", "0.726", "-0.025", "-0.061", "-0.042", "0.007"],
  ["0.932", "0.841", "-0.101", "-0.074", "0.011", "0.057"]
]
-- Verify we have a good approximate answer:


/**/ indent([ [ FloatStr(eval(F, pt)) | pt In Sol ]
 | F in System]);
[
  ["0.0000", "0.0000"],
  ["0.0000", "0.0000"],
  ["0.0000", "0.0000"],
  ["-4.6479*10^(-78)", "-1.2625*10^(-80)"],
  ["3.2795*10^(-78)", "3.0111*10^(-80)"],
  ["-3.9126*10^(-78)", "-1.9934*10^(-80)"]
]
```

In other words we have that the squares of the linear forms

$$\ell_1 = x + (-0.030..)y + (0.726..)z \quad \text{and} \quad \ell_2 = x + (0.932..)y + (0.841..)z$$

are in the ideal generated by q_1, q_2, q_3, q_4 in $\mathbb{R}[x, y, z]$, and more precisely

$$\ell_1^2 = (-0.025..)q_1 + (-0.061..)q_2 + (-0.042..)q_3 + (0.007..)q_4$$
$$\ell_2^2 = (-0.101..)q_1 + (-0.074..)q_2 + (0.011..)q_3 + (0.057..)q_4$$

In conclusion we have shown that $I \subset \mathbb{Q}[x, y, z]$ does not contain the square of a linear form with coefficients in $\mathbb{Q}$, whereas the ideal $(q_1, q_2, q_3, q_4) \subset \mathbb{R}[x, y, z]$ does. This means, by Proposition 63, that $\mathbb{Q}[x, y, z]/(q_1, q_2, q_3, q_4)$ is not G-quadratic, whereas $\mathbb{R}[x, y, z]/(q_1, q_2, q_3, q_4)$ might be.

Exercise 67 Consider the ideal generated by t random quadrics in n indeterminates. Choose t, n and look for square of linear forms in the ideal.

9 Koszul Algebras in a Nutshell

In the previous sections we have introduced various notions related to Koszulness, and counterexamples showing that they are not equivalent.

In the next page we summarize into a handy diagram all results and counterexamples we have shown in this chapter.

resolution of I 2-linear

$25 \Downarrow$

R artinian $\Rightarrow \ell^2 \in I$ $\mathrm{gin}_{\mathrm{DegRevLex}}(I)$
generated by quadrics

Prop 63 $\nwarrow$ *Rem* 17 $\Downarrow$ $\not\Uparrow$ *Ex* 19

R has Gröbner flag   *Prop* 57 $\Rightarrow$
$\not\Leftarrow$
Remark 58 R is G-quadratic

$\not\Uparrow$
Ex 64   *trivial* $\Downarrow$ $\not\Uparrow$ *Ex* 49

I generated
by c.i. of quadrics   *Prop* 61 $\Rightarrow$ R is LG-quadratic

Rem 46 $\Downarrow$ $\not\Uparrow$ *Ex* 54

R has Koszul filtration

Lemma 53 $\searrow$
$\not\nwarrow$
Remark 58 $\boxed{\textbf{R is Koszul}}$
$\nearrow$
Prop 36

$P_K^R(t) \cdot H_R(-t) = 1$ *Rem* 31 $\Downarrow$ $\not\Uparrow$ *Ex* 37

Rem 11 $\Downarrow$ R is quadratic

$1/H_R(-t)$ has
no negative coefficients

Primary Decompositions

with Sections on Macaulay2 and Networks

Irena Swanson and Eduardo Sáenz-de-Cabezón

Abstract This chapter contains three major sections, each one roughly corresponding to a lecture. The first section is on computing primary decompositions, the second one is more specifically on binomial ideals, and the last one is on some primary decomposition questions in algebraic statistics and networks.

1 Computation of Primary Decompositions

In a polynomial ring in one variable, say $R = \mathbb{Q}[x]$, it is easy to compute the primary decomposition say of $(x^4 - 1)$:

$$(x^4 - 1) = (x^2 + 1) \cap (x - 1) \cap (x + 1).$$

The reason that this computation is easy is that we readily found the irreducible factors of the polynomial $x^4 - 1$. In general, finding irreducible factors is a necessary prerequisite for the computation of primary decompositions. In these notes we make the STANDING ASSUMPTION that for any field k that arises as a finite field extension of $\mathbb{Q}$ or of a finite field, and for any variable x over k, one can compute all irreducible factors of any polynomial in $k[x]$. The reader interested in more details about polynomial factorization should consult [116] or [69, p. 38].

Throughout all rings are Noetherian and commutative with identity.

I. Swanson (✉)
Reed College, 3203 SE Woodstock Blvd, Portland, OR 97202, USA
e-mail: iswanson@reed.edu

E. Sáenz-de-Cabezón
Departamento de Matemáticas y Computación, Universidad de La Rioja, c/ Madre de Dios 53,
26006 Logroño, La Rioja, Spain
e-mail: eduardo.saenz-de-cabezon@unirioja.es

© Springer International Publishing AG 2017

A.M. Bigatti et al. (eds.), *Computations and Combinatorics in Commutative Algebra*, Lecture Notes in Mathematics 2176, DOI 10.1007/978-3-319-51319-5_2

1.1 Introduction to Primary Ideals and Primary Decompositions

Definition 1 An ideal I in a ring R is **primary** if $I \neq R$ and every zerodivisor in R/I is nilpotent.

Facts

1. *Any prime ideal is primary.*
2. *If I is a primary ideal, then $\sqrt{I} = \{r \in R : r^l \in I \text{ for some } l \in \mathbb{N}\}$ is a prime ideal. Furthermore, if $P = \sqrt{I}$, then I is also called P-**primary**.*
3. *If I is P-primary, there exists a positive integer n such that $P^n \subseteq I$.*
4. *The intersection of any two P-primary ideals is P-primary.*
5. *If $\sqrt{I}$ is a prime ideal, it need not be the case that I is primary, nor is it the case that the square of a prime ideal is primary. For example, let P be the kernel of the ring homomorphism $k[X, Y, Z] \rightarrow k[t]$ taking X to t^3, Y to t^4, and Z to t^5. Then $P = (x^3 - yz, y^2 - xz, z^2 - x^2y)$ is a prime ideal, the radical of P^2 is $P, x^5 + xy^3 - 3x^2yz + z^3 \notin P^2$ by an easy degree count, $x \notin P$, but*

$$x(x^5 + xy^3 - 3x^2yz + z^3) = (x^3 - yz)^2 - (y^2 - xz)(z^2 - x^2y),$$

which proves that P^2 is not primary.
6. *Suppose that I is an ideal such that $\sqrt{I}$ is a maximal ideal. Then I is a primary ideal. Namely, if $r \in R$ is a zerodivisor modulo I, then as R/I is Artinian with only one maximal ideal, necessarily the image of r is in this maximal ideal. But then a power of r lies in I.*
7. *Let P be a prime ideal and I a P-primary ideal. Then for any $r \in R$,*

$$I : r = \begin{cases} I, & \text{if } r \notin P \\ R, & \text{if } r \in I \\ a \ P - \text{primary ideal strictly containing } I, & \text{if } r \in P \setminus I. \end{cases}$$

Moreover, there exists $r \in R$ such that $I : r = P$.
8. *Let $R \rightarrow S$ be a ring homomorphism, and I a primary ideal in S. Then $I \cap R$ is primary to $\sqrt{I} \cap R$.*
9. *Let U be a multiplicatively closed subset of R. There is a one-to-one correspondence between prime (resp. primary) ideals in R disjoint from U and prime (resp. primary) ideals in $U^{-1}R$ given by $I \mapsto IU^{-1}R$ for I an ideal in R, and $J \mapsto J \cap R$ for J an ideal in $U^{-1}R$.*
10. *If I is P-primary and x is a variable over R, then $IR[x]$ is $PR[x]$-primary.*

Definition 3 Let I be an ideal in a ring R. A decomposition $I = \cap_{i=1}^{s} q_i$ is a **primary decomposition** of I if $q_1, \ldots, q_s$ are primary ideals.

If in addition all $\sqrt{q_i}$ are distinct and for all i, $\cap_{j \neq i} q_j \not\subseteq q_i$, then the decomposition is called **irredundant** or **minimal**.

By Facts 2, the following is immediate:

Proposition 4 *If* $I = \cap_{i=1}^{s} q_i$ *is a (minimal) primary decomposition, then for any multiplicatively closed set* U *such that* $U^{-1}I \neq U^{-1}R$,

$$U^{-1}I = \bigcap_{q_i \cap U = \emptyset} U^{-1} q_i$$

is a minimal primary decomposition. □

Emmy Noether proved the existence of primary decompositions:

Theorem 5 *Every proper ideal* I *in a Noetherian ring* R *has a (minimal) primary decomposition.*

Proof Once existence of a primary decomposition is established, existence of a minimal one is straightforward: if the radicals of two components are identical, we replace the two components with one component, namely their intersection, and if one component contains the intersection of the others, then that one component is redundant and is omitted. So it suffices to prove the existence of any primary decomposition.

If I is primary, the decomposition consists of I only. In particular, if I is a maximal ideal, it has a primary decomposition. So assume that I is not primary. Then by definition there exist $a, b \in R$ such that $ab \in I, a \notin I$ and $b \notin \sqrt{I}$. As R is Noetherian, the chain $I \subseteq I : b \subseteq I : b^2 \subseteq \cdots$ terminates. Choose n such that $I : b^l = I : b^{l+1} = \cdots$. It is straightforward to prove that $I = (I : b^l) \cap (I + (b^l))$. By assumption $a \in (I : b^l) \setminus I$ and $b^l \in (I + (b^l)) \setminus I$. Thus both $I : b^l$ and $I + (b^l)$ properly contain I. By Noetherian induction, these two larger ideals have a primary decomposition, and the intersection of the two decompositions gives a possibly redundant primary decomposition of I. □

Observe that the proof above is rather non-constructive: how does one decide whether an ideal is primary, and even if somehow one knows that an ideal is not primary, how can one determine the elements a and b? Nevertheless, this is a crucial step in the algorithm for computing primary decompositions in polynomial rings that we present. An important point for algorithmic computing is also that the ascending chain $I \subseteq I : b \subseteq I : b^2 \subseteq \cdots$ is special: as soon as we have one equality $I : b^l = I : b^{l+1}$, then for all $m \geq l, I : b^l = I : b^m$. (General ascending chains do not have this property.)

Example 6 For monomial ideals it is straightforward to decide when they are primary: a monomial ideal I in $R = k[X_1, \ldots, X_n]$ is primary if and only if whenever a variable X_j divides some minimal monomial generator of I, then a power of X_j is contained in I. This fact at the same time makes the existence of primary decompositions of monomial ideals, as outlined in the proof of Theorem 5, constructive. Namely, it is easy to check if each factor of each minimal monomial generator has a power in I. If yes, the ideal is primary, otherwise there exists a monomial generator a with variable $b = X_j$ dividing a and $b \notin \sqrt{I}$. We then repeat the construction as in the proof of Theorem 5 to obtain two strictly larger monomial ideals, and use Noetherian induction. In particular, we apply this to $I = (x^2, xy, xz)$. With $b = y$ and $a = x$ we get that $I : y = I : y^2$ and so that

$$I = (I : y) \cap (I + (y)) = (x) \cap (x^2, y, xz).$$

Now (x) is already primary (even prime), but (x^2, y, xz) is not. We apply the proof of Theorem 5 with $b = z, a = x$ to get that $(x^2, y, xz) = ((x^2, y, xz) : z) \cap ((x^2, y, xz) + (z)) = (x, y) \cap (x^2, y, z)$, so that

$$I = (x) \cap (x, y) \cap (x^2, y, z).$$

Clearly (x, y) is redundant, so that finally we get the minimal primary decomposition

$$I = (x) \cap (x^2, y, z).$$

But this is not the only possible primary decomposition. Namely, in the last step we could have used $(x^2, y, xz) = ((x^2, y, xz) : z^2) \cap ((x^2, y, xz) + (z^2)) = (x, y) \cap (x^2, y, xz, z^2)$, to get that

$$I = (x) \cap (x, y) \cap (x^2, y, xz, z^2) = (x) \cap (x^2, y, xz, z^2),$$

which gives a different primary decomposition.

This gives an example of non-uniqueness of primary decompositions. However, certain uniqueness does hold:

Theorem 7 *If* $I = q_1 \cap \cdots \cap q_s$ *is a minimal primary decomposition, then* $\{\sqrt{q_1}, \ldots, \sqrt{q_s}\}$ *equals the set of all prime ideals of the form* $I : f$ *as* f *varies over elements of* R. *In particular, the set* $\{\sqrt{q_1}, \ldots, \sqrt{q_s}\}$ *is uniquely determined. If* $\sqrt{q_i}$ *is minimal (under inclusion) in this set, then* q_i *is uniquely determined as*

$$I_{\sqrt{q_i}} \cap R.$$

More generally, for each i, *there exists* $l_i \in \mathbb{N}$ *such that* $\sqrt{q_i}^{l_i} \subseteq q_i$. *Then*

$$I = \bigcap_{i=1}^{s} \left((\sqrt{q_i}^{l_i} + I)_{\sqrt{q_i}} \cap R \right)$$

is also a primary decomposition.

Proof By minimality of the primary decomposition, for each i there exists $r \in \cap_{j \neq i} q_j \setminus q_i$. Then $I : r = (q_1 : r) \cap \cdots \cap (q_s : r) = q_i : r$ is primary to $\sqrt{q_i}$, and by Facts 2, there exists $r' \in R$ such that $q_i : (rr') = (q_i : r) : r'$ equals $\sqrt{q_i}$. Conversely, suppose that $I : f$ is a prime ideal. This means that $(q_1 : f) \cap \cdots \cap (q_s : f)$ is a prime ideal, so necessarily this prime ideal equals some $q_i : f$. But by Facts 2, necessarily this prime ideal equals $\sqrt{q_i}$. This proves the first two statements of the theorem.

The third statement follows from Facts 2 and Proposition 4, and the fourth one from Facts 2. For the last statement, observe that $(\sqrt{q_i}^{l_i} + I)_{\sqrt{q_i}}$ is primary to the maximal ideal and contained in the localization of q_i, so that $(\sqrt{q_i}^{l_i} + I)_{\sqrt{q_i}} \cap R$ is $\sqrt{q_i}$-primary and contained in q_i. Since it also contains I, it follows that

$$I \subseteq \bigcap_{i=1}^{s} \left(\sqrt{q_i}^{l_i} + I \right)_{\sqrt{q_i}} \cap R \right) \subseteq \bigcap_{i=1}^{s} q_i = I,$$

so that equality holds throughout.   □

The primes appearing in this theorem are called **associated primes**, and their set is denoted as $\mathrm{Ass}(R/I)$. When the l_i are taken to be minimal possible, the resulting primary decomposition is called **canonical** (see works by Ortiz [90], Ojeda and Piedra-Sánchez [88, 89] and Ojeda [87]).

Yao proved that the (non-unique) primary components can be mixed and matched more generally than in the last statement in the theorem:

Theorem 8 ("Mix-and-match", Yao [118]) *Let* $\{P_1, \ldots, P_s\} = \mathrm{Ass}(R/I)$, *and assume that for $j = 1, \ldots, s$,*

$$I = \bigcap_{i=1}^{s} q_{ji},$$

is a primary decomposition of I with $\sqrt{q_{ji}} = P_i$ *for all i, j. Then $I = \bigcap_{i=1}^{s} q_{ii}$ is also a primary decomposition.*

The following appeared in the proof of Theorem 5: for any element $b \in R$ and any ideal I of R, $I \subseteq I : b \subseteq I : b^2 \subseteq \cdots$. By Noetherian assumption, there exists l such that $I : b^l = I : b^{l+1}$, and hence $I = (I : b^l) \cap (I + (b^l))$. Thus straightforwardly

$$\mathrm{Ass}\left(\frac{R}{I : b^l} \right) \subseteq \mathrm{Ass}\left(\frac{R}{I} \right) \subseteq \mathrm{Ass}\left(\frac{R}{I : b^l} \right) \bigcup \mathrm{Ass}\left(\frac{R}{I + (b^l)} \right).$$

Incidentally, the stable value of $I : b^n$ is also often written as $I : b^\infty$.

It is left as an exercise that

$$\mathrm{Ass}\left(\frac{R}{I : b} \right) \subseteq \mathrm{Ass}\left(\frac{R}{I} \right) \subseteq \mathrm{Ass}\left(\frac{R}{I : b} \right) \bigcup \mathrm{Ass}\left(\frac{R}{I + (b)} \right)$$

even when $I \neq (I : b) \cap (I + (b))$. This latter fact can be very helpful for example if b is a variable, so that a primary decomposition of $I + (b)$ is essentially done in the polynomial ring in fewer variables and can thus possibly be handled by induction on the dimension of the polynomial ring.

By Noetherian induction we know all the associated primes of $I : b, I : b^l, I + (b)$ and $I + (b^l)$. By the two set inclusions displayed above, all the associated primes of $I : b$ and $I : b^l$ are associated to I. In general, not all associated primes of $I + (b)$ and $I + (b^l)$ are associated to I. Thus the two displays above generate sets of prime ideals that include all the possible associated primes of I, but with possible redundancies. The following result can help resolve the redundancies:

Proposition 9 *A prime ideal P is associated to an ideal I if and only if P is minimal over $I : (I : P^\infty)$.*

Proof Both parts are preserved under localization at P, so we may assume that the ring is local with P being the maximal ideal. Then $I : P^\infty$ is the intersection of all primary components of I that are not P-primary, so that $I : (I : P^\infty)$ is either the ring if P is not associated, and is a P-primary ideal otherwise. $\qquad\square$

We also leave as an exercise the useful fact that if I is homogeneous in a $\mathbb{Z}^d$-graded ring, then so are all of the associated primes of I, and there exists a primary decomposition of I all of whose components are homogeneous. This has to do with zerodivisors in graded rings.

1.2 Computing Radicals and Primary Decompositions

In this section we present the Gianni-Trager-Zacharias algorithm [60]. We use Gröbner bases and induction on the number of variables. By the STANDING ASSUMPTION we can compute radicals and primary decompositions in $k[X_1, \ldots, X_n]$ if $n \leq 1$. Now suppose that $n > 1$.

Alternate algorithms for computing primary decompositions can be found in the paper [52] by Eisenbud et al. and in the paper [102] by Shimoyama and Yokoyama. A survey with clear exposition on algorithms and the current state of computation is in the paper [42] by Decker et al.

Reduction Step 1

Proposition 10 *Let $A = k[X_1, \ldots, X_d] \subseteq R = k[X_1, \ldots, X_n]$ where k is a field. Then for any ideal I in R, $I_{A\backslash(X_1)} \cap R$ is computable.*

Proof The proof shows how to compute it.

We impose the lexicographic order $X_n > \cdots > X_1$ on R. Any term t in R can be written as aM_t, where a is a term in A and M_t is a monomial in $k[X_{d+1}, \ldots, X_n]$. For each $f \in R$, let $\widetilde{f}$ be the sum of all those terms t in f for which $M_t = M_{\text{ltf}}$. Write $\widetilde{f} = a_f X_1^{e_f} M_{\text{ltf}}$ for some non-negative integer e_f and some $a_f \in A \backslash (X_1)$. We also write M_f for M_{ltf}.

Let G be a Gröbner basis of I.

Claim If $f \in I_{A\backslash(X_1)} \cap R$ then there exist $g \in G$ and $r \in A \backslash (X_1)$ such that $\widetilde{rf} \in \widetilde{g}R$.

Proof of the claim Let $f \in I_{A\setminus(X_1)} \cap R$. Then for some $c \in A \setminus (X_1), cf \in I$, so that $\mathrm{lt}(cf)$ is a multiple of $\mathrm{lt}g$ for some $g \in G$. Write $\mathrm{lt}(cf) = aX_1^e M(\mathrm{lt}g)$ for some $a \in A \setminus (X_1), e \in \mathbb{N}$, and some monomial M in $k[X_{d+1}, \ldots, X_n]$. We will prove that it is possible to find g such that $e_{cf} \geq e + e_g$. Suppose that $e_{cf} < e + e_g$. Then there exists a term in cf that is a $k[X_2, \ldots, X_d]$-multiple of $X_1^{e_{cf}} M_{cf}$ and that is not cancelled in $cf - aX_1^e Mg$. Thus $cf - aX_1^e Mg$ has a term t with $M_t = M_{cf}$ and $e_t = e_{cf} < e + e_g$. Suppose that we have $a_1, \ldots, a_{s-1} \in A \setminus (X_1), M_1, \ldots, M_{s-1}$ monomials in $k[X_{d+1}, \ldots, X_n]$, and non-negative integers $e_1, \ldots, e_{s-1}$ such that for all $j = 1, \ldots, s - 1, \mathrm{lt}(cf - \sum_{i=1}^{j-1} a_i X_1^{e_i} M_i g_i) = \mathrm{lt}(a_j X_1^{e_j} M_j g_j)$, and $e_{cf} < e_{g_j} + e_j$. Set $h = cf - \sum_{i=1}^{s-1} a_i X_1^{e_i} M_i g_i$. By the last conditions, $M_h = M_{cf} = M_j M_{g_j}$ for all j. As h is in I, we have that the initial term of h is $a_s X_1^{e_s} M_s(\mathrm{lt}g_s)$ for some $g_s \in G, a_s \in A \setminus (X_1), e_s \in \mathbb{N}$, and some monomial M_s in $k[X_{d+1}, \ldots, X_n]$. Since the monomial ordering is a well-ordering, this cannot go on forever, so that for some $g \in G, e_{cf} \geq e_g + e$. But then $a_g \widetilde{cf} = a_g \widetilde{cf} = a_f X_1^{e_{cf} - e_g} M \widetilde{g}$. This proves the claim.

Set $b = \prod_{g \in G} a_g$. Certainly $I_b \cap R \subseteq I_{A\setminus(X_1)} \cap R$. Now let $f \in I_{A\setminus(X_1)} \cap R$. To prove that $f \in I_b \cap R$, it suffices to assume that among all f in $(I_{A\setminus(X_1)} \cap R) \setminus I_b$, the term M_f is smallest. By the claim, there exist $g \in G, r \in A \setminus (X_1)$ and $h \in R$ such that $\widetilde{rf} = h\widetilde{g} = ha_g X_1^{e_g} M_g$. Let $u = \gcd(r, h)$. Then $\frac{r}{u} f = \frac{h}{u} a_g X_1^{e_g} M_g$. Since R is a UFD, necessarily $\frac{r}{u} \in A \setminus (X_1)$ is a factor of a_g, hence of b. Write $b = v\frac{r}{u}$. Then $bf = v\frac{r}{u} f = v\frac{h}{u} a_g X_1^{e_g} M_g$. Set $h = bf - v\frac{h}{u} g$. By construction, $M_h < M_{bf} = M_f$. If $M_f = 1$, then $h = 0$, and in general, $h \in I_{A\setminus(p)} \cap R$. By induction on $M_h, h \in I_b \cap R$, so that $bf = h + v\frac{h}{u} g \in I_b \cap R$, whence $f \in I_b \cap R$. This proves that $I_b \cap R = I_{A\setminus(p)} \cap R$.

Finally, $I_b \cap R = I : b^\infty$ is computable because $I : b^\infty$ is the first stabilization in the inclusions $I \subseteq I : b \subseteq I : b^2 \subseteq I : b^3 \subseteq \cdots$. $\qquad\square$

Reduction Step 2 To compute a primary decomposition, we reduce to the case where $I \cap A$ is primary for all subrings A of R generated over k by a proper subset of the variables $X_1, \ldots, X_n$.

Proof Fix one such A. Let $J = I \cap A$. By induction we can compute a minimal primary decomposition $J = q_1 \cap \cdots \cap q_s$. If $s = 1$, we are done, so we suppose that $s > 1$. We want to identify i such that $\sqrt{q_i}$ is a minimal associated prime ideal. We want to accomplish this with minimal computing effort. We could certainly compute all the radical ideals and compare them, but computing radicals can be time-consuming, so the radical is not a goal in itself, we avoid its computation. Instead, we compute some colon ideals. If $q_1 : q_i \neq q_1$ for some $i > 1$, then $\sqrt{q_1}$ is definitely not a minimal prime, so we can eliminate q_1 from further pairwise tests. If instead $q_1 : q_i = q_1$ for all $i = 2, \ldots, s$, then $\sqrt{q_1}$ is a minimal prime ideal. With such cloning, in finitely many steps we identify i such that $\sqrt{q_i}$ is a minimal prime ideal. Say $i = 1$.

Now we want $r \in q_2 \cap \cdots \cap q_s \setminus \sqrt{q_1}$. Certainly we can find an element $r \in q_2 \cap \cdots \cap q_s$ but avoiding q_1 as follows: one of the generators of $q_2 \cap \cdots \cap q_s$ is not in q_1, and this can be tested. By prime avoidance, it is even true that a random/generic element r of $q_2 \cap \cdots \cap q_s$ is not in $\sqrt{q_1}$. Ask the computer to give you a random element r of $q_2 \cap \cdots \cap q_s$, and then $r \notin \sqrt{q_1}$ if and only if $q_1 : r = q_1$. Thus while random generation may not reliably produce an element of $q_2 \cap \cdots \cap q_s \setminus \sqrt{q_1}$, we do have a computable method via colon of checking for this property. In practice, one would probably ask for one random r, test it, and if the test fails, ask for a new random element, and if necessary repeat a small finite number of times. A reader uncomfortable with the randomness of this procedure, should instead compute $\sqrt{q_1}$, and then test successively for a generator of $q_2 \cap \cdots \cap q_s$ to not be in $\sqrt{q_1}$.

So suppose that we have $r \in q_2 \cap \cdots \cap q_s \setminus \sqrt{q_1}$. As on page 45, there exists a positive integer l such that $I : r^l = I : r^{l+1}$. This ideal is strictly larger than I as it contains $q_1 R$. Furthermore, $I + (r^l)$ is strictly larger than I since $r \notin \sqrt{q_1}$ and hence $r \notin \sqrt{I}$. If we can obtain a primary decomposition of the strictly larger ideals $I : r^l$ and $I + (r^l)$, then we get one also for $I = (I : r^l) \cap (I + (r^l))$. Thus by replacing I by the strictly larger ideals $I : r^l$ and $I + (r^l)$, we get strictly larger intersections with A, and we continue this until the intersections are primary.

We repeat this procedure with all the possible A. While working on a new $I \cap A'$, the intersections $I \cap A$ with the old A can only get larger, but by the Noetherian property of A it can get larger only finitely many times. Since there are only finitely many possible A this procedure has to stop.

Reduction Step 2 To compute a primary decomposition, we reduce to the case where $I \cap k[X_i]$ is non-zero for all i.

Suppose that $I \cap k[X_1] = (0)$. This is a principal prime ideal, so that by Proposition 10, there is a computable non-zero $b \in k[X_1]$ such that $Ik(X_1)[X_2, \ldots, X_n] \cap R = I : b^\infty$. Let l be a (computable) positive integer such that $I : b^\infty = I : b^l$. The ideal $I + (b^l)$ has the desired property that its intersection with $k[X_1]$ is not zero. Since $I = (I : b^l) \cap (I + (b^l))$, it suffices to find a primary decomposition of $I : b^l$.

By induction on the number of variables, we can compute a minimal primary decomposition $Ik(X_1)[X_2, \ldots, X_n] = q_1 \cap \cdots \cap q_s$. If $s = 1$, then by the one-to-one correspondence between primary ideals before and after localization, $I : b^l$ is primary, and we are done. So we may assume that $s > 1$. Then as in the proof of Reduction step 1 we can compute $r \in k(X_1)[X_2, \ldots, X_n]$ that is a non-nilpotent zerodivisor modulo $Ik(X_1)[X_2, \ldots, X_n]$. We can write $r = \frac{r_1}{r_2}$ for some $r_1 \in R, r_2 \in A \setminus (X_1)$, and by ignoring the unit r_2 we may assume that $r = r_1 \in R$. Then I is the intersection of strictly larger ideals $I : r^l$ and $I + (r^l)$ in R, and we proceed by Noetherian induction on ideals in R.

We repeat this with $I \cap k[X_i]$ for all $i > 1$.

Reduction Step 3 To compute a primary decomposition, we reduce to the case where $I \cap k[X_i]$ is non-zero for all i and $I \cap A$ is primary for all subrings A of R generated over k by a proper subset of the variables $X_1, \ldots, X_n$.

For this repeat the first two reduction steps. Again by Noetherian induction in each of the finitely many rings this step terminates in finitely many steps.

Reduction Step 4 To compute the radical, we reduce to the case where $I \cap k[X_i]$ is non-zero for all i and $I \cap A$ is primary for all subrings A of R generated over k by a proper subset of the variables $X_1, \ldots, X_n$.

Note that Reduction step 1 for the computation of primary decompositions successively replaces I by strictly larger ideals $J_1, \ldots, J_s$ such that $I = J_1 \cap \cdots \cap J_s$ and such that $J_i \cap A$ is primary for all A and all i. Since $\sqrt{I} = \sqrt{J_1} \cap \cdots \cap \sqrt{J_s}$, it suffices to compute $\sqrt{J_i}$ for all i.

If $I \cap k[X_1] = (0)$, by induction on the number of variables we can compute the radical of $Ik(X_1)[X_2, \ldots, X_n]$. Let $g_1, \ldots, g_t$ be a generating set of this radical. By possibly clearing denominators, we may assume that $g_1, \ldots, g_t \in R$. Then the radical of $Ik(X_1)[X_2, \ldots X_n]$ intersected with R equals $J = (g_1, \ldots, g_t)k(X_1)[X_2, \ldots X_n] \cap R$. This is a radical ideal, and it is computable by Proposition 10. Certainly $\sqrt{I} \subseteq J$. More precisely by Proposition 10, there exists non-zero $b \in k[X_1]$ such that $(g_1, \ldots, g_t)k(X_1)[X_2, \ldots X_n] \cap R = (g_1, \ldots, g_t) : b^\infty$. Then $I : b^\infty = I : b^l$ for some l, $I = (I : b^l) \cap (I + (b^l))$, and the radical of I is $J \cap \sqrt{I + (b^l)}$, so it suffices to compute the radical of the strictly larger ideal $I + (b^l)$. So we may assume that $I \cap k[X_1] \neq (0)$, and more generally that $I \cap k[X_i] \neq (0)$ for all i.

Repetition of this and Noetherian induction bring to a successful reduction in this step.

Theorem 11 *The radical and the primary decomposition of an ideal I in R are computable.*

Proof We have reduced to the case where $I \cap k[X_1] = (f_1), \ldots, I \cap k[X_n] = (f_n)$, and $I \cap k[X_1, \ldots, X_{n-1}]$ are primary.

By our STANDING ASSUMPTION, $(p_i) = \sqrt{(f_i)}$ is computable. In characteristic zero, this computation is easier: $p_i = \frac{f_i}{\gcd(f_i, f_i')}$.

By induction on the number of variables we can compute the radical of $I \cap k[X_1, \ldots, X_{n-1}]$. Since we assumed that $I \cap k[X_1, \ldots, X_{n-1}]$ is primary, it follows that its radical is a maximal ideal; call it M. (In characteristic zero, as in [70], $M = I \cap k[X_1, \ldots, X_{n-1}] + (p_1, \ldots, p_{n-1})$ because $k[X_1, \ldots, X_{n-1}]/(p_1, \ldots, p_{n-1}) = (k[X_1]/(p_1)) \otimes_k \cdots \otimes_k (k[X_{n-1}]/(p_{n-1}))$ is a tensor product of finitely generated field extensions of k, and is thus reduced, semisimple, so that any ideal in this ring is radical.)

Since $I \cap k[X_n] \neq (0)$, necessarily I is not a subset of MR. We can compute $g \in I \setminus MR$. Even more, since $R/MR = \frac{k[X_1, \ldots, X_{n-1}]}{M}[X_n]$ is a principal ideal domain, we can compute $g \in I$ such that $g(R/MR) = I(R/MR)$. By the STANDING ASSUMPTION, there exists $g_1, \ldots, g_s \in R$ such that the $g_i(R/MR)$ are pairwise non-associated and irreducible, and such that $g(R/MR) = g_1^{a_1} \cdots g_s^{a_s}(R/MR)$ for some positive integers $a_1, \ldots, a_s$.

Then $I \subseteq \cap_i(MR + g_iR) = MR + (g_1 \cdots g_s)R \subseteq \sqrt{I}$, the associated primes of I are $MR + g_iR, i = 1, \ldots, s$, $\sqrt{I} = \cap_i(MR + g_iR)$, and the $(MR + g_iR)$-primary component of I is $I : (\prod_{j \neq i} g_j)^\infty$. All of these are computable. $\qquad \square$

Example 12 Let $I = (x^2 + yz, xz - y^2, x^2 - z^2)$ in $\mathbb{Q}[x, y, z]$. We roughly follow the outline of the algorithm, with some human ingenuity to skip computational steps. Clearly $yz + z^2 \in I \cap k[y, z]$ and it appears unlikely that a power of z is contained in $I \cap k[y,z]$. (We could use elimination and Gröbner bases to compute precisely $I \cap k[y,z] = (yz+z^2, y^3+z^3)$.) Thus z is a non-nilpotent zerodivisor modulo I. By the algorithm we compute $I : z = (y+z, xz-z^2, x^2-z^2), I : z^2 = (y+z, x-z) = I : z^3$, which is clearly prime and hence primary. Furthermore, $I+(z^2) = (x^2, yz, xz-y^2, z^2)$ has radical (x,y,z), which is a maximal ideal, so that $I + (z^2)$ is primary. Thus $I = (I : z^2) \cap (I + (z^2)) = (y+z, x-z) \cap (x^2, yz, xz-y^2, z^2)$ is a primary decomposition, and clearly it is an irredundant one.

1.3   Computer Experiments: Using Macaulay2 to Obtain Primary Decompositions

The computer algebra system Macaulay2 [62] has in-built functions to deal with primary decompositions. There is a package, included with the system, that is devoted to this topic. In this section we encourage the reader to turn on the computer, start a Macaulay2 session and experiment with the software.

To see the capabilities of Macaulay2 with respect to primary decompositions, one can first read the help pages for the package. One can do this in two ways: typing help PrimaryDecomposition in the command line interface, or reading the html version in a browser (by typing viewHelp in the command line interface. We rapidly review the main functions Macaulay2 offers to compute primary decompositions.

The first thing to do is of course typing your favourite ideal and using the in-built function primaryDecomposition:

```
i1 : R=QQ[x,y,z];

i2 : I=ideal(x^2,x*y,x*z);

o2 : Ideal of R

i3 : primaryDecomposition I

                          2
o3 = {ideal(x), ideal (x , y, z)}
```

We can immediately obtain the associated primes of I (in the order corresponding to the primary components):

```
i4 : associatedPrimes I

o4 = {ideal(x), ideal (x, y, z)}
```

This is because when computing the primary decomposition, Macaulay2 caches the information it obtains, which can be accessed at any time, without further computations:

```
i1 : R=QQ[x,y,z];

i2 : I=ideal(x^3,x*y,x*z);

o2 : Ideal of R

i3 : peek I.cache

o3 = CacheTable{}

i4 : primaryDecomposition I

                  3
o4 = {ideal(x), ideal (x , y, z)}

i6 : peek I.cache

o6 = CacheTable{AssociatedPrimes => {ideal(x), ideal (x, y, z)}
          module => image | x3 xy xz |

        flattenRing => OptionTable{CoefficientRing => null}
                    3
        => (ideal (x , x*y, x*z), map(R,R,{x, y, z}))
                            Result =>(Thing, RingMap)
```

Macaulay2 is able to use different algorithms to compute primary decompositions; they are called **strategies** in the system. They are sensitive to the input ideal:

```
i1 : R=QQ[x,y,z];

i2 : I=ideal(x^3+y+1,y^3+z+1,z^3+x+1);

o2 : Ideal of R

i3 : J=I^2;

o3 : Ideal of R

i4 : K=I^2;

o4 : Ideal of R

i5 : L=I^2;

o5 : Ideal of R

i6 : time primaryDecomposition J;
     -- used 1.27953 seconds
```

```
i7 : time primaryDecomposition (K, Strategy=>
     EisenbudHunekeVasconcelos);
     -- used 49.3968 seconds
```

```
i8 : time primaryDecomposition (L, Strategy=>
     new Hybrid from (1,2));
     -- used 41.828 seconds
```

```
i9 : peek J.cache
```

```
i10: peek K.cache
```

```
i11: peek L.cache
```

Note that the output of lines i9, i10 and i11 is too long to be printed here. We encourage the reader to check it in her/his own computer. The cached information makes a difference when obtaining further information about the ideal. The algorithms available for computing primary decompositions are Shimoyama and Yokoyama [102], Eisenbud et al. [52], a hybrid of these two algorithms, and Gianni et al. [60]. The default algorithm in Macaulay2 is Shimoyama-Yokoyama. Macaulay2 has also special strategies for monomial and binomial ideals.

2 Expanded Lectures on Binomial Ideals

In these pages I present the commutative algebra gist of the Eisenbud–Sturmfels paper [51]. The paper employs lattice and character theory, but this presentation, inspired by Melvin Hochster's, avoids that machinery.

The main results are that the associated primes, the primary components, and the radical of a binomial ideal in a polynomial ring are binomial if the base ring is algebraically closed.

Kahle wrote a program [68] that computes binomial decompositions extremely fast: the input fields do not have to be algebraically closed, but the program adds the necessary roots of numbers.

Throughout, $R = k[X_1, \ldots, X_n]$, where k is a field and $X_1, \ldots, X_n$ are variables over k. A **monomial** is an element of the form $\underline{X}^{\underline{a}}$ for some $a \in \mathbb{N}_0^n$, and a **term** is a scalar multiple of a monomial. The words "monomial" and "term" are often confused, and in particular, a **binomial** is defined as the difference of two terms. (In my opinion, we should switch the meanings of "monomial" and "term".) An ideal is **binomial** if it is generated by binomials.

Here are some easy facts:

1. Every monomial is a binomial, hence every monomial ideal is a binomial ideal.
2. The sum of two binomial ideals is a binomial ideal.
3. The intersection of binomial ideals need not be binomial: $(t - 1) \cap (t - 2) = t^2 - 3t + 2$, which is not binomial in characteristics other than 2 and 3.

4. Primary components of a binomial ideal need not be binomial: in $\mathbb{R}[t]$, the binomial ideal (t^3-1) has exactly two primary components: $(t-1)$ and (t^2+t+1).
5. The radical of a binomial ideal need not be binomial: Let t, X, Y be variables over $\mathbb{Z}/2\mathbb{Z}, k = (\mathbb{Z}/2\mathbb{Z})(t), R = k[X, Y]$, and $I = (X^2 + t, Y^2 + t + 1)$. Note that I is binomial (as $t + 1$ is in k), and $\sqrt{I} = (X^2 + t, X + Y + 1)$, and this cannot be rewritten as a binomial ideal as there is only one generator of degree 1 and it is not binomial.

Thus, for the announced good properties of binomial ideals, we do need to make a further assumption, namely, **from now on**, all fields k are algebraically closed, and then the counterexamples to primary components and radicals do not occur.

Can the theory be extended to trinomial ideals (with obvious meanings)? The question is somewhat meaningless, because **all ideals are trinomial** after adding variables and a change of variable. Namely, let $f = a_1 + a_2 + \cdots + a_m$ be a polynomial with m terms. Introduce new variables $t_3, \ldots, t_m$. Then $k[X_1, \ldots, X_n]/(f) = k[X_1, \ldots, X_n, t_3, \ldots, t_m]/(a_1 + a_2 - t_3, -t_3 + a_3 - t_4, -t_4 + a_4 - t_5, \ldots, -t_{m-2} + a_{m-2} - t_{m-1}, -t_{m-1} + a_{m-1} - t_m)$. In this way an ideal I in a polynomial ring can be rewritten for some purposes as a trinomial ideal in a strictly larger polynomial ring, so that essentially every ideal is trinomial in this sense. Then the general primary decomposition and radical properties follow-after adding more variables.

But binomial ideals are special. By Buchberger's algorithm, a Gröbner basis of a binomial ideal is binomial: all S-polynomials and all reductions of binomial ideals with respect to binomials are binomial. Thus whenever I is a binomial ideal and A is a polynomial subring generated by some of the variables of R, then $I \cap A$ is binomial. In particular, from the commutative algebra fact that $I \cap J = (tI + (t-1)J)R[t] \cap R$, where t is a variable over R, whenever I is binomial and J is monomial, then $I \cap J$ is binomial. Similarly, for any monomial j, $I \cap (j)$ and $I : j$ are binomial.

Proposition 13 *Let I be a binomial ideal, and let $J_1, \ldots, J_l$ be monomial ideals. Then there exists a monomial ideal J such that $(I + J_1) \cap \cdots \cap (I + J_l) = I + J$.*

Proof We can take a k-basis B of R/I to consist of monomials. By Gröbner bases of binomial ideals, $(I + J_k)/I$ is a subspace whose basis is a subset of B. Thus $\cap((I + J_k)/I)$ is a subspace whose basis is a subset of B, which proves the proposition. $\square$

Binomial ideals are sensitive to the coefficients appearing in the generators. This has implications in complexity theory, as well as in practical computations. For example, if the characteristic of k is not 0 and R is a polynomial ring in $m \times n$ variables X_{ij}, the ideal generated by the 2×2-determinants of $[X_{ij}]_{i,j}$ is a prime ideal (see for example [30]), whereas the ideal generated by such permanents (both coefficients $+1$) generate a prime ideal precisely when $m = n = 2$, they generate a radical ideal precisely when $\min\{m, n\} \leq 2$, and whenever $m, n \geq 3$, the number of minimal primes is $n + m + \binom{n}{2}\binom{m}{2}$. (This is due to [73].)

2.1 Binomial Ideals in
$$S = k[X_1, \ldots, X_n, X_1^{-1}, \ldots, X_n^{-1}] = k[X_1, \ldots, X_n]_{X_1 \cdots X_n}$$

Any binomial $\underline{X}^{\underline{a}} - c\underline{X}^{\underline{b}}$ can be written up to unit in S as $\underline{X}^{\underline{a}-\underline{b}} - c$.

Let I be a proper binomial ideal in S. Write $I = (\underline{X}^{\underline{e}} - c : \underline{e} \in \mathbb{Z}^n, c_e \in k^*)$. (All c_e are non-zero since I is assumed to be proper.)

If e, e' occur in the definition of I, set $e'' = e - e', e''' = e + e'$. Then

$$\underline{X}^{\underline{e}} - c_e = \underline{X}^{\underline{e}'+\underline{e}''} - c_e \equiv c_{e'}\underline{X}^{\underline{e}''} - c_e \mod I,$$

$$\underline{X}^{\underline{e}} - c_e = \underline{X}^{\underline{e}'''-\underline{e}'} - c_e \equiv c_{e'}^{-1}\underline{X}^{\underline{e}'''} - c_e \mod I,$$

so that e'' is allowed with $c_{e''} = c_e c_{e'}^{-1}$, and e''' is allowed with $c_{e'''} = c_e c_{e'}$. In particular, the set of all allowed e forms a $\mathbb{Z}$-submodule of $\mathbb{Z}^n$. Say that it is generated by m vectors. Record these vectors into an $n \times m$ matrix A. We just performed some column reductions: neither these nor the rest of the standard column reductions over $\mathbb{Z}$ change the ideal I. But we can also perform column reductions! Namely, S also equals $k[X_1 X_2^m, X_2, \ldots, X_n, (X_1 X_2^m)^{-1}, (X_2)^{-1}, \ldots, (X_n)^{-1}]$, and we can rewrite any monomial $\underline{X}^{\underline{a}}$ as $(X_1 X_2^m)^{a_1} X_2^{a_2-ma_1} X_3^{a_3} \cdots X_n^{a_n}$, which corresponds to the second row of the matrix becoming the old second row minus m times the old first row (and other rows remain unchanged). Simultaneously we changed the variables, but not the ring. So all row reductions are allowed, they do not change the ideal, but they do change the ideal. We work this out on an example:

Example 14 Let $I = (x^3 y - 7y^3 z, xy - 4z^2)$ in $k[x, y, z]$, where the characteristic of k is different from 2 and 7. This yields the 3×2 matrix of occurring exponents:

$$A = \begin{bmatrix} 3 & 1 \\ -2 & 1 \\ -1 & -2 \end{bmatrix}.$$

We will keep track of the coefficients 7 and 4 for the columns like so:

$$A = \begin{bmatrix} 3 & 1 \\ -2 & 1 \\ -1 & -2 \end{bmatrix}$$
$$7 \quad 4$$

We first perform some elementary column reductions, keeping track of the c_e (if all c_e are 1, then there is no reason to keep track of these, they will always be 1):

$$A \to \begin{bmatrix} 1 & 3 \\ 1 & -2 \\ -2 & -1 \end{bmatrix} \to \begin{bmatrix} 1 & 0 \\ 1 & -5 \\ -2 & 5 \end{bmatrix}$$
$$7 \quad 4 \qquad\qquad 4 \quad 7/4^3$$

We next perform the row reductions, and for these we will keep track of the names of variables (in the obvious way):

$$
\begin{matrix} x \\ y \\ z \end{matrix}
\begin{bmatrix} 1 & 0 \\ 1 & -5 \\ -2 & 5 \end{bmatrix}
\rightarrow
\begin{matrix} xy \\ y \\ z \end{matrix}
\begin{bmatrix} 1 & 0 \\ 0 & -5 \\ -2 & 5 \end{bmatrix}
\rightarrow
\begin{matrix} xyz^{-2} \\ y \\ z \end{matrix}
\begin{bmatrix} 1 & 0 \\ 0 & -5 \\ 0 & 5 \end{bmatrix}
\rightarrow
\begin{matrix} xyz^{-2} \\ y \\ zy^{-1} \end{matrix}
\begin{bmatrix} 1 & 0 \\ 0 & 0 \\ 0 & 5 \end{bmatrix}
\rightarrow
\begin{matrix} xyz^{-2} \\ zy^{-1} \\ y \end{matrix}
\begin{bmatrix} 1 & 0 \\ 0 & 5 \\ 0 & 0 \end{bmatrix}.
$$

In these reductions, the coefficients remained 4 and $7/4^3$.

This was only a special case, but obviously the procedure works for any binomial ideal in S: the matrix A can be row- and column-reduced, keeping track of the variables and coefficients. Once we bring the matrix of exponents into standard form, every proper binomial ideal in S is of the form $((X_1)^{\prime m_1} - c_1, \ldots, (X_d)^{\prime m_d} - c_d)$ for some $d \le n$, some $m_i \in \mathbb{N}$, some $c_i \in K^*$, and some X_i' are products of positive and negative powers of $X_1, \ldots, X_n$ in a way that keeps the ring equality $S = k[X_1', \ldots, X_n', X_1'^{-1}, \ldots, X_n'^{-1}]$.

Now the following are obvious: in characteristic zero,

$$
I = \bigcap_{u_i^{m_i} = c_i} (X_1' - u_1, \ldots, X_d' - u_d),
$$

where all the primary components are distinct, binomial, and prime. Thus here all associated primes, all primary components, and the radical are all binomial ideals, and moreover all the associated primes have the same height and are thus all minimal over I.

In positive prime characteristic p, write each m_i as $p^{v_i} n_i$ for some positive v_i and non-negative n_i that is not a multiple of p. Then

$$
I = \bigcap_{u_i^{m_i} = c_i} ((X_1' - u_1)^{p^{v_1}}, \ldots, (X_d' - u_d)^{p^{v_d}}).
$$

The listed generators of each component are primary. These primary components are binomial, as $(X_i' - u_i)^{p^{v_i}} = X_i'^{p^{v_i}} - u_i^{p^{v_i}}$. The radicals of these components are all the associated primes of I, and they are clearly the binomial ideals $(X_1' - u_1, \ldots, X_d' - u_d)$. All of these prime ideals have the same height, thus they are all minimal over I. Furthermore,

$$
\sqrt{I} = \bigcap_{u_i^{m_i} = c_i} (X_1' - u_1, \ldots, X_d' - u_d) = (X_1'^{n_1} - u_1^{n_1}, \ldots, X_d'^{n_d} - u_d^{n_d}),
$$

for any u_i with $u_i^{m_i} = c_i$. The last equality is in fact well-defined as if $(u_i')^{m_i} = c_i$, then $0 = c_i - c_i = u_i^{m_i} - (u_i')^{m_i} = (u_i^{n_i} - (u_i')^{n_i})^{p^{v_i}}$, so that $u_i^{n_i} = (u_i')^{n_i}$. In particular, $\sqrt{I}$ is binomial.

We summarize this section in the following theorem:

Theorem 15 *A proper binomial ideal in S has binomial associated primes, binomial primary components, and binomial radical. All associated primes are minimal. In characteristic zero, all components are prime ideals, so all binomial ideals in S are radical. In positive prime characteristic p, a generating set of a primary component consists of (different) Frobenius powers of the elements in some binomial generating set of the corresponding prime ideal.*   □

Example 16 In particular, if we analyze the ideal from Example 14, the already established row reduction shows that $I = (xyz^{-2} - 4, (zy^{-1})^5 - 7/4^3)$. In characteristic 5, this is a primary ideal whose radical is $I = (xyz^{-2} - 4, zy^{-1} - \sqrt[5]{7/4^3}) = (xyz^{-2} - 4, zy^{-1} - 3) = (xy - 4z^2, z - 3y) = (xy - 4 \cdot 9y^2, z - 3y) = (x - y, z - 3y)$. In characteristics other than $2, 5, 7$, we get five associated primes $(xy - 4z^2, z - \alpha y) = (x - 4\alpha^2 y, z - \alpha y)$ as α varies over the fifth roots of $7/4^3$. All of these prime ideals are also the primary components of I. (In characteristics 2 and 7, $IS = S$.)

Theorem 17 *Let I be an ideal in R such that IS is binomial. Then $IS \cap R$ is binomial. In particular, for any binomial ideal I of R, any associated prime ideal P of I such that $PS \neq S$ is binomial, and we may take the P-primary component of I (in R) to be binomial.*

Proof Let Q be a binomial ideal in R such that $QS = IS$. Then $IS \cap R = QS \cap R = Q : (X_1 \cdots X_n)^\infty$ is binomial by the facts at the beginning of this section.   □

2.2 Associated Primes of Binomial Ideals Are Binomial

Theorem 18 *All associated primes of a binomial ideal are binomial ideals. (Recall that k is algebraically closed.)*

Proof By factorization in polynomial rings in one variable, the theorem holds if $n \leq 1$. So we may assume that $n \geq 2$. The theorem is clearly true if the binomial ideal I is a maximal ideal. Now let I be arbitrary.

Let $j \in [n] = \{1, \ldots, n\}$. Note that $I + (X_j) = I_j + (X_j)$ for some binomial ideal I_j in $k[X_1, \ldots, X_{n-1}]$. By induction on n, all prime ideals in $\mathrm{Ass}(k[X_1, \ldots, X_{n-1}]/I_j)$ are binomial. But $\mathrm{Ass}(R/(I + (X_j))) = \{P + (X_j) : P \in \mathrm{Ass}(k[X_1, \ldots, X_{n-1}]/I_j)\}$, so that all prime ideals in $\mathrm{Ass}(R/(I + (X_j)))$ are binomial. By the basic facts from the beginning of this section, $I : X_j$ is binomial. If X_j is a zerodivisor modulo I, then $I : X_j$ is strictly larger than I, so that by Noetherian induction, $\mathrm{Ass}(R/(I : X_j))$ contains only binomial ideals. By facts on page 45, $\mathrm{Ass}(R/I) \subseteq \mathrm{Ass}(R/(I+(X_j))) \cup \mathrm{Ass}(R/(I : X_j))$, whence also by induction on the number of variables, all associated primes of I are binomial as long as some variable is a zerodivisor modulo I.

Now assume that all variables are non-zerodivisors modulo I. Let $P \in \mathrm{Ass}(R/I)$. Since $X_1 \cdots X_n$ is a non-zerodivisor modulo I, it follows that $P_{X_1 \cdots X_n} \in \mathrm{Ass}((R/I)_{X_1 \cdots X_n}) = \mathrm{Ass}(S/IS)$. Then P is binomial by Theorem 17.   □

We have already seen in Example 6 that for monomial ideals all associated primes are monomial (hence binomial).

Example 19 (Continuation of Examples 14 and 16) Let $I = (x^3y - 7y^3z, xy - 4z^2)$ in $k[x, y, z]$. We have already determined all associated prime ideals of I that do not contain any variables. So it suffices to find the associated primes of $I + (x^m), I + (y^m)$ and of $I + (z^m)$, for large m. If the characteristic of k is 2, then the decomposition is

$$I = (x^3y - y^3z, xy)I = (y^3z, xy) = (y) \cap (y^3, x) \cap (z, x),$$

If the characteristic of k is 7, then the decomposition is

$$I = (x^3y, xy - 4z^2) = (y, z^2) \cap (x^3, xy - 4z^2).$$

(The reader may apply methods of the previous section to verify that the latter ideal is primary.) Now we assume that the characteristic of k is different from 2 and 7. Any prime ideal that contains I and x also contains z, so at least we have that (x, z) is minimal over I and thus associated to I. Similarly, (y, z) is minimal over I and thus associated to I. Also, any prime ideal that contains I and z contains in addition either x or y, so that at least we have determined $\text{Min}(R/I)$. Any embedded prime ideal would have to contain all of the already determined primes. Since I is homogeneous, all associated primes are homogeneous, and in particular, the only embedded prime could be (x, y, z). It turns out that this prime ideal is not associated even if it came up in our construction, but we won't get to this until we discuss the theory of primary decomposition of binomial ideals in the next section.

2.3 Primary Decomposition of Binomial Ideals

The main goal of this section is to prove that every binomial ideal has a binomial primary decomposition, if the underlying field is algebraically closed (Theorem 23). We first need a lemma and more terms.

Definition 20 An ideal I in a polynomial ring $k[X_1, \ldots, X_n]$ is **cellular** if for all $i = 1, \ldots, n, X_i$ is either a non-zerodivisor or nilpotent modulo I.

All primary monomial and binomial ideals are cellular.

Definition 21 For any binomial $g = \underline{X}^{\underline{a}} - c\underline{X}^{\underline{b}}$ and for any non-negative integer d, define

$$g^{[d]} = \underline{X}^{d\underline{a}} - c^d\underline{X}^{d\underline{b}}.$$

The following is a crucial lemma:

Lemma 22 *Let I be a binomial ideal, let $g = \underline{X}^{\underline{a}} - c\underline{X}^{\underline{b}}$ be a non-monomial binomial in R such that $\underline{X}^{\underline{a}}$ and $\underline{X}^{\underline{b}}$ are non-zerodivisors modulo I. Then there exists a monomial ideal I_0 such that for all large d, $I : g^{[d!]} = I : (g^{[d!]})^2 = I + I_0$.*

Proof For all positive integers d and e, $g^{[d]}$ is a factor of $g^{[de]}$, so that $I : g^{[d]} \subseteq I : g^{[de]}$. Thus there exists d such that for all $e \geq d$, $I : g^{[d!]} = I : g^{[e!]}$.

Let $f \in I : g^{[d!]}$. Write $f = f_1 + f_2 + \cdots + f_s$ for some terms (coefficient times monomial) $f_1 > f_2 > \cdots > f_s$. Without loss of generality $\underline{X}^{\underline{a}} > \underline{X}^{\underline{b}}$. We have that

$$f_1\underline{X}^{\underline{a}} + f_2\underline{X}^{\underline{a}} + \cdots + f_s\underline{X}^{\underline{a}} + f_1\underline{X}^{\underline{b}} + f_2\underline{X}^{\underline{b}} + \cdots + f_s\underline{X}^{\underline{b}} \in I.$$

In the Gröbner basis sense, each $f_i\underline{X}^{\underline{a}}, f_i\underline{X}^{\underline{b}}$ reduces to some unique term (coefficient times monomial) modulo I. Since $\underline{X}^{\underline{a}}$ and is a non-zerodivisor modulo I, $f_i\underline{X}^{\underline{a}}$ and $f_j\underline{X}^{\underline{a}}$ cannot reduce to a scalar multiple of the same monomial, and similarly $f_i\underline{X}^{\underline{b}}$ and $f_j\underline{X}^{\underline{b}}$ cannot reduce to a scalar multiple of the same monomial. Thus for each $j = 1, \ldots, s$ there exists $\pi(j) \in [s] = \{1, \ldots, s\}$ such that $f_j\underline{x}^{d!\underline{a}} - c^{d!}f_{\pi(j)}\underline{x}^{d!\underline{b}} \in I$. The function $\pi : [s] \to [s]$ is injective. By easy induction, for all i, $f_j(\underline{x}^{d!\underline{a}})^i - c^{d!i}f_{\pi^i(j)}(\underline{x}^{d!\underline{b}})^i \in I$. By elementary group theory, $\pi^{s!}(j) = j$, so that for all j, $f_jg^{[d!][s!]} \in I$. Then $f_jg^{[((d!)(s!))!]} \in I$, and by the choice of d, $f_jg^{[d!]} \in I$. Thus $I : g^{[d!]}$ contains monomials $f_1, \ldots, f_s$. Thus set I_0 to be the monomial ideal generated by all the monomials appearing in the generators of $I : g^{[d!]}$.

Let $f \in I : (g^{[d!]})^2$. We wish to prove that $f \in I : g^{[d!]}$. By possibly enlarging I_0 we may assume that I_0 contains all monomials in $I : g^{[d!]} = I + I_0$. This in particular means that any Gröbner basis G of $I : g^{[d!]}$ consists of monomials in I_0 and binomial non-monomials in I. Write $f = f_1 + f_2 + \cdots + f_s$ for some terms $f_1 > f_2 > \cdots > f_s$. As in the previous paragraph, for each j, either $f_j\underline{x}^{d!\underline{a}} \in I_0$ or else $f_j\underline{x}^{d!\underline{a}} - c^{d!}f_{\pi(j)}\underline{x}^{d!\underline{b}} \in I$. If $f_j\underline{x}^{d!\underline{a}} \in I_0 \subseteq I : g^{[d]}$, then by the non-zerodivisor assumption, $f_j \in I : g^{[d]}$, which contradicts the assumption. So necessarily we get the injective function $\pi : [s] \to [s]$. As in the previous paragraph we then get that each $f_j \in I : g^{[d]}$. $\square$

Without loss of generality assume that no f_i is in $I : g^{[d!]}$. Note that $fg^{[d!]} \in I : g^{[d!]}$. Consider the case that $f_j\underline{x}^{d!\underline{a}} \in I_0$ and get a contradiction. Now repeat the π argument as in a previous part to make the conclusion.

Theorem 23 *If k is algebraically closed, then any binomial ideal has a binomial primary decomposition.*

Proof Let I be a binomial ideal. For each variable X_j there exists l such that $I = (I : X_j^l) \cap (I + (X_j)^l)$, so it suffices to find the primary decompositions of the two ideals $I : X_j^l$ and $I + (X_j)^l$. These two ideals are binomial, the former by the basic facts from the beginning of this section. By repeating this splitting for another X_i on each of the two new ideals, and then repeating for X_k on the four new ideals, et cetera, with even some j repeated, we may assume that each of the intersectands is cellular.

Thus it suffices to prove that each cellular binomial ideal has a binomial primary decomposition.

So let I be cellular and binomial. By possibly reindexing, we may assume that $X_1, \ldots, X_d$ are non-zerodivisors modulo I, and $X_{d+1}, \ldots, X_n$ are nilpotent modulo I. Let $P \in \mathrm{Ass}(R/I)$. By Theorem 18, P is a binomial prime ideal. Since I is contained in P, P must contain $X_{d+1}, \ldots, X_n$, and since the other variables are non-zerodivisors modulo I, these are the only variables in P. Thus $P = P_0 + (X_{d+1}, \ldots, X_n)$, where P_0 is a binomial prime ideal whose generators are binomials in $k[X_1, \ldots, X_d]$, and $X_1, \ldots, X_d$ are non-zerodivisors modulo I.

So far we have I "cellular with respect to variables". (For example, we could have $I = (X_3(X_1^2 - X_2^2), X_3^2)$ and $P = (X_1 - X_2, X_3)$.) Now we will make it more "cellular with respect to binomials in the subring". Namely, let g be a non-zero binomial in P_0. (In the parenthetical example, we could have $g = X_1 - X_2$.) By Lemma 22, there exists $d \in \mathbb{N}$ such that $I : g^{[d]} = I : (g^{[d]})^2 = I +$ (monomial ideal). This in particular implies that P is not associated to $I : g^{[d]}$, and so necessarily P is associated to $I + (g^{[d]})$. Furthermore, the P-primary component of I is the P-primary component of the binomial ideal $I + (g^{[d]})$. We replace the old I by the larger binomial ideal $I + (g^{[d]})$. We repeat this to each g a binomial generator of P_0, so that we may assume that P is minimal over I. (In the parenthetical example above, we would now say with $d = 6$ that $I = (X_1^6 - X_2^6, X_3(X_1^2 - X_2^2), X_3^2)$.) Now $X_{d+1}, \ldots, X_n$ are still nilpotent modulo I. The P-primary component of I is the same as the P-primary component of binomial ideal $I : (X_1 \cdots X_d)^\infty$, so by replacing I with $I : (X_1 \cdots X_d)^\infty$ we may assume that I is still cellular.

If $\mathrm{Ass}(R/I) = \{P\}$, then I is P-primary, and we are done. So we may assume that there exists an associated prime ideal Q of I different from P. Since P is minimal over I and different from Q, necessarily there exists an irreducible binomial $g = \underline{X}^{\underline{a}} - c\underline{X}^{\underline{b}} \in Q \setminus P$. Necessarily $g \notin (X_{d+1}, \ldots, X_n)R$. Thus Lemma 22 applies, so there exists $d \in \mathbb{N}$ such that $I : g^{[d]} = I : (g^{[d]})^2 = I +$ (monomial ideal). Note that Q is not associated to this ideal but Q is associated to I, so that the binomial ideal $I : g^{[d]}$ is strictly larger than I. If $g^{[d]} \notin P$, then the P-primary component of I equals the P-primary component of $I : g^{[d]}$, and so by Noetherian induction (if we have proved it for all larger ideals, we can then prove it for one of the smaller ideals) we have that the P-primary component of I is binomial. So without loss of generality we may assume that $g^{[d]} \in P$. Then $g^{[d]}$ contains a factor in P of the form $g_0 = \underline{X}^{\underline{a}} - c'\underline{X}^{\underline{b}}$ for some $c' \in k$. If the characteristic of R is p, $g_0^{p^m}$ is a binomial for all m, we choose the largest m such that p^m divides d, and set $h = g^{[d]}/g_0, b = g_0^{p^m}$. In characteristic zero, we set $h = g^{[d]}/g_0$ and $b = g_0$. In either case, b is a binomial, $b \in I : h$ and $h \notin P$. Thus the P-primary component of I is the same as the P-primary component of $I : h$, and in particular, since $I \subseteq I + (b) \subseteq I : h$, it follows that the P-primary component of I is the same as the P-primary component of the binomial ideal $I + (b)$. If $b \in Q$, then $g_0 = \underline{X}^{\underline{a}} - c'\underline{X}^{\underline{b}}$ and $g = \underline{X}^{\underline{a}} - c\underline{X}^{\underline{b}}$ are both in Q. Necessarily $c \neq c'$, so that $\underline{X}^{\underline{a}}, \underline{X}^{\underline{b}} \in Q$, and since $g \notin (X_{d+1}, \ldots, X_n)R$, it follows that Q contains one of the variables $X_1, \ldots, X_d$. But these variables are non-zerodivisor modulo I, so that Q cannot be associated to I, which proves that

$b \notin Q$. But then I is strictly contained in $I + (b)$, and by Noetherian induction, the P-primary component is binomial. $\qquad\square$

2.4   The Radical of a Binomial Ideal Is Binomial

Here is general commutative algebra fact: for any Noetherian commutative ring R, any ideal I, and any $X_1, \ldots, X_n$ in R,

$$\sqrt{I} = \sqrt{I + (X_1)} \cap \cdots \cap \sqrt{I + (X_n)} \cap \sqrt{I : (X_1 \cdots X_n)^\infty}$$
$$= \sqrt{I + (X_1)} \cap \cdots \cap \sqrt{I + (X_n)} \cap \sqrt{I : X_1 \cdots X_n}.$$

Theorem 24 *The radical of any binomial ideal in a polynomial ring over an algebraically closed field is binomial.*

Proof This is clear if $n = 0$. So assume that $n > 0$. By the fact above,

$$\sqrt{I} = \sqrt{I + (X_1)} \cap \cdots \cap \sqrt{I + (X_n)} \cap \sqrt{I : (X_1 \cdots X_n)^\infty}.$$

Let $I_0 = \sqrt{I : (X_1 \cdots X_n)^\infty}$. We have established in Theorem 15 that $\sqrt{I_0}S = \sqrt{IS}$ is binomial in S. By Theorem 17, $\sqrt{I_0}$ is binomial.

Let $I_1 = I \cap k[X_2, \ldots, X_n]$. We know that I_1 is binomial. By induction on n, the radical of I_1 is binomial. This radical is contained in $\sqrt{I}$, so that $\sqrt{I} = \sqrt{\sqrt{I_1} + I}$. Thus without loss of generality we may assume that $\sqrt{I_1} \subseteq I$. Hence we may also assume that $\sqrt{I_1} = I_1$.

Let P be a prime ideal minimal over $I + (X_1)$. Suppose that there exists a binomial g in I that involves X_1 but is not in (X_1). Write $g = X_1 m' + m$ for some monomial terms m, m', with X_1 not appearing in m. Since P is a prime ideal, there exists a variable dividing m that is in P. Say this variable is X_2. Then P is a prime ideal minimal over $I + (X_1, X_2)$. By continuing this we get that, after reindexing, P is a prime ideal minimal over $I + (X_1, X_2, \ldots, X_d)$ and that any binomial in I is either in $(X_1, \ldots, X_d)$ or in $k[X_{d+1}, \ldots, X_d]$. By Gröbner bases rewriting,

$$I + (X_1, \ldots, X_d) = ((I + (X_1, \ldots, X_d)) \cap k[X_{d+1}, \ldots, X_n] + (X_1, \ldots, X_d))R$$
$$= (I_1 \cap k[X_{d+1}, \ldots, X_n] + (X_1, \ldots, X_d))R,$$

and this is a radical ideal since I_1 is. This proves that the intersection of all the prime ideals minimal over $I + (X_1)$ equals the intersection of ideals of the form $I +$ (some variables). Hence by Proposition 13, $\sqrt{I + (X_1)} = I + J_1$ for some monomial ideal J_1. Similarly, $\sqrt{I + (X_i)} = I + J_i$ for some monomial ideals $J_1, \ldots, J_n$. By the first paragraph in this section and by Proposition 13 then $\sqrt{I} = (I + J) \cap I_0$ for some monomial ideal J. But $I \subseteq I_0$, so that $\sqrt{I} = I + J \cap I_0$, and this is a binomial ideal because J is monomial and I_0 is binomial (see p. 53). $\qquad\square$

3   Primary Decomposition in Algebraic Statistics

Algebraic statistics is a relatively new field. The first systematic work is due to
Studený [109] from an axiomatic point of view, and several works after that used
the axiomatic approach. A first more concrete connection between statistics and
commutative algebra is due to the paper of Diaconis and Sturmfels [46], which
introduced the notion of a Markov basis. The book by Pistone et al. [92], published
in 2001, is a book on commutative algebra and Gröbner bases for statisticians.
Not all parts of statistics can be algebraicized, of course. Some of the current
research topics in algebraic statistics are: design of experiments, graphical models,
phylogenetic invariants, parametric inference, maximum likelihood estimation,
applications in biology, et cetera. This section is about (conditional) independence.

3.1   Conditional Independence

Definition 25 A **random variable**, as used in probability and statistics, is not a
variable in the algebra sense; it is a variable or function whose value is subject to
variations due to chance. I cannot give a proper definition of "chance", but let us
just say that examples of random variables are outcomes of flips of coins or rolls of
dice. (If you are Persi Diaconis, a flip of a coin has a predetermined outcome, but
not if I flip it.)

A **discrete random variable** is a random variable that can take on at most finitely
many values (such as the flip of a coin or the roll of a die).

Throughout we will be using the standard notation $P(i)$ to stand for the
probability that condition i is satisfied, and $P(i \mid j)$ to stand for the **(conditional)**
probability that condition i is satisfied given that condition j holds. Whenever
$P(j) \neq 0$, then

$$P(i \mid j) = \frac{P(i, j)}{P(j)}.$$

Definition 26 Random variables Y_1, Y_2 are **independent** for all possible values i of
Y_1 and all possible values j of Y_2, $P(Y_1 = i \mid Y_2 = j) = P(Y_1 = i)$, or in other
words, if

$$P(Y_1 = i, Y_2 = j) = P(Y_1 = i)P(Y_2 = j).$$

If this is satisfied, we write $Y_1 \perp\!\!\!\perp Y_2$.

Let $p_{ij} = P(Y_1 = i, Y_2 = j)$. Then $\sum_j p_{ij} = P(Y_1 = i)$ and $\sum_i p_{ij} = P(Y_2 = j)$.
(In statistics, these sums are shortened to p_{i+} and p_{+j}, respectively.) For discrete
random variables Y_1, Y_2, with Y_1 taking on all values in $[m]$ and Y_2 taking on all

values in $[n]$, independence is equivalent to the following matrix equality:

$$
\begin{bmatrix} p_{11} & \cdots & p_{1n} \\ \vdots & & \vdots \\ p_{m1} & \cdots & p_{mn} \end{bmatrix} = \begin{bmatrix} P(Y_1 = 1) \\ P(Y_1 = 2) \\ \vdots \end{bmatrix} \begin{bmatrix} P(Y_2 = 1) \cdots P(Y_2 = n) \end{bmatrix}.
$$

Since the sum of the p_{ij} is 1, it follows that the rank of the matrix $[p_{ij}]$ is 1, and so $I_2([p_{ij}]) = 0$. Conversely, if $I_2([p_{ij}]) = 0$, since some p_{ij} is non-zero, necessarily $[p_{ij}]$ has rank 1. Then we can write

$$
\begin{bmatrix} p_{11} & \cdots & p_{1n} \\ \vdots & & \vdots \\ p_{m1} & \cdots & p_{mn} \end{bmatrix} = \begin{bmatrix} a_1 \\ a_2 \\ \vdots \\ a_m \end{bmatrix} \begin{bmatrix} b_1 \cdots b_2 \cdots b_m \end{bmatrix}
$$

for some real numbers a_i, b_j. Since some p_{ij} is a positive real number, by possibly multiplying all a_i and b_j by -1 we may assume that all a_i, b_j are non-negative real numbers. Let $a = \sum_i a_i, b = \sum_j b_j$. Then

$$
ab = \sum_{i,j} a_i b_j = \sum_{i,j} p_{ij} = 1,
$$

whence we also have

$$
\begin{bmatrix} p_{11} & \cdots & p_{1n} \\ \vdots & & \vdots \\ p_{m1} & \cdots & p_{mn} \end{bmatrix} = \begin{bmatrix} a_1 b \\ \vdots \\ a_m b \end{bmatrix} \begin{bmatrix} ab_1 \cdots ab_2 \cdots ab_m \end{bmatrix}.
$$

All the entries of the two matrices on above are non-negative, $a_i b = \sum_j a_i b_j = \sum_j p_{ij} = P(Y_1 = i)$ and $ab_j = \sum_i a_i b_j = \sum_i p_{ij} = P(Y_2 = j)$, which yields the factorization of $[p_{ij}]$ as in the rephrasing of independence. Thus $Y_1 \perp\!\!\!\perp Y_2$ if and only if $I_2([p_{ij}]_{i,j}) = 0$.

How does one decide independence in practice? Say a poll counts people according to their hair length and whether they watch soccer as follows:

	Watches soccer	Does not watch soccer
Has short hair	400	200
Has long hair	40	460

Thus watching soccer and the hair length in this group appear to not be independent: it seems that the hair length fairly determines whether one watches soccer. Even if the polling has a 10% error in representing the population, it still seems that the hair length fairly determines whether one watches soccer. However, the poll break-down among genders shows the following:

Men	Watching	*Not*
Short hair	400	100
Long hair	40	10

Women	Watching	Not
Short hair	0	100
Long hair	0	450

Now, given the gender, the probability that one watches soccer is independent of hair length (odds for watching is 4/5 for men, 0 for women).

This brings up an issue: in general one does not find such clean numbers with determinant precisely 0, and so one has to do further manipulations of the data to decide whether it is statistically likely that there is an independence of data.

Here I continue with the obvious needed definition arising from the previous example:

Definition 27 Random variables Y_1 and Y_2 are **independent given** the random variable Y_3, if for every value i of Y_1, j of Y_2 and k of Y_3,

$$P(Y_1 = i \mid Y_2 = j, Y_3 = k) = P(Y_1 = i \mid Y_3 = k).$$

If $P(Y_3 = k) > 0$, this is equivalent to saying that $P(Y_1 = i, Y_2 = j, Y_3 = k)P(Y_3 = k) = P(Y_1 = i)P(Y_2 = j)$. We write such independence as $Y_1 \perp\!\!\!\perp Y_2 \mid Y_3$.

Let M be the 3-dimensional hypermatrix whose (i,j,k) entry is $P(Y_1 = i, Y_2 = j, Y_3 = k)$, $Y_1 \perp\!\!\!\perp Y_2 \mid Y_3$. Then $Y_1 \perp\!\!\!\perp Y_2 \mid Y_3$ means that on each k-level of M, the ideal generated by the 2×2-minors of the matrix on that level is 0.

Here are the **axioms** of conditional independence:

1. **Triviality:** $X \perp\!\!\!\perp \emptyset \mid Z$. (Algebraically this says that the ideal generated by the 2×2-minors of an empty matrix is 0.)
2. **Symmetry:** $X \perp\!\!\!\perp Y \mid Z$ implies $Y \perp\!\!\!\perp X \mid Z$. (Algebraically this follows as the ideal of minors of a matrix as the same as the ideal of the transpose of that matrix.)
3. **Weak union:** $X \perp\!\!\!\perp \{Y_1, Y_2\} \mid Z$ implies $X \perp\!\!\!\perp Y_1 \mid \{Y_2, Z\}$. Here we point out that if U and V is a (discrete) random variable, so is $\{U, V\}$, whose values are pairs of values of U and V, of course. (Algebraically this says the following: let $p_{ijkl} = P(X = i, Y_1 = j, Y_2 = k, Z = l)$. The assumption says that for all values l of Z, the ideal generated by the 2×2-minors of the matrix $[p_{ijkl}]_{i,(j,k)}$ is 0. But then for fixed l and a fixed value k of Y_2, the ideal generated by the 2×2-minors of the submatrix $[p_{ijkl}]_{i,j}$ is 0 as well, which is the conclusion.)
4. **Decomposition:** $X \perp\!\!\!\perp \{Y_1, Y_2\} \mid Z$ implies $X \perp\!\!\!\perp Y_1 \mid Z$. (Algebraically this says that if for each l, $I_2([p_{ijkl}]_{i,(j,k)}) = 0$ then $I_2([p_{ij+l}]_{i,j}) = 0$, where $+$ means that the corresponding entry is the sum $\sum_k p_{ijkl}$.)

5. **Contraction:** $X \perp\!\!\!\perp Y \mid \{Z_1, Z_2\}$ and $X \perp\!\!\!\perp Z_2 \mid Z_1$ implies $X \perp\!\!\!\perp \{Y, Z_2\} \mid Z_1$. (Algebraically this says that if for each $k, l, I_2([p_{ijkl}]_{i,j}) = 0$ and for each $k, I_2([p_{i+kl}]_{i,l}) = 0$, then for each $k, I_2([p_{ijkl}]_{i,(j,l)}) = 0$.)
6. **Intersection axiom:** Under the assumption that all $P(X = i, Y = j, Z = k)$ are positive, $X \perp\!\!\!\perp Y \mid Z$ and $X \perp\!\!\!\perp Z \mid Y$ implies $X \perp\!\!\!\perp \{Y, Z\}$.

The last axiom is the focus of the next section.

3.2 Intersection Axiom

Algebraically the intersection axiom says that if all p_{ijk} are positive, if for each $k, I_2([p_{ijk}]_{i,j}) = 0$, and if for each $j, I_2([p_{ijk}]_{i,k}) = 0$, then $I_2([p_{ijk}]_{i,(j,k)}) = 0$.

Example 28 Here we show that the assumption on the p_{ijk} being positive is necessary. Let M be the $2 \times 2 \times 2$-hypermatrix whose (i, j, k) entry is

$$p_{ijk} = \begin{cases} 1/8, & \text{if } i = j = k = 1; \\ 3/8, & \text{if } i = 2, j = k = 1; \\ 3/8, & \text{if } i = 1, j = k = 2; \\ 1/8, & \text{if } i = j = k = 2; \\ 0, & \text{otherwise.} \end{cases}$$

We can view this in a $2 \times 2 \times 2$-hypermatrix, with the third axis going up, the second axis going to the right, and the first axis coming out of the page:

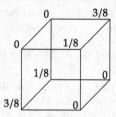

Then

$$[p_{ij1}]_{i,j} = \begin{bmatrix} 1/8 & 0 \\ 3/8 & 0 \end{bmatrix}, \quad [p_{ij2}]_{i,j} = \begin{bmatrix} 0 & 3/8 \\ 0 & 1/8 \end{bmatrix},$$

$$[p_{i1k}]_{i,k} = \begin{bmatrix} 1/8 & 3/8 \\ 0 & 0 \end{bmatrix}, \quad [p_{i2k}]_{i,k} = \begin{bmatrix} 0 & 0 \\ 3/8 & 1/8 \end{bmatrix},$$

and all have zero determinants. However,

$$[p_{ijk}]_{i,(j,k)} = \begin{bmatrix} 1/8 \; 0 \; 0 \; 3/8 \\ 3/8 \; 0 \; 0 \; 1/8 \end{bmatrix}$$

in which one 2×2-minor is not zero. Note that the last matrix is the flattening of the hypermatrix—squish into the $x - y$ plane, without any overlaps.

The intersection axiom says that if all p_{ijk} are non-zero, the conditions on the vanishing on the minors along each k and along each j-level are enough to make the "slanted" 2×2-minors zero as well.

We parse the intersection axiom further. Now let X_{ijk} stand for a variable (algebraic, not random, variable). The axiom says that the simultaneous zero $\underline{\alpha}$ of $I_2([X_{ijk}]_{i,j})$ for each k and of $I_2([X_{ijk}]_{i,k})$ for each j is also a zero of $I_2([X_{ijk}]_{i,(j,k)})$ if all entries in $\underline{\alpha}$ are positive. Via Hilbert's Nullstellensatz this says that

$$I_2([X_{ijk}]_{i,(j,k)}) \subseteq \sqrt{\sum_k I_2([X_{ijk}]_{i,j}) + \sum_j I_2([X_{ijk}]_{i,k}) : (\prod_{i,j,k} X_{ijk})^{\infty}}.$$

Certainly

$$\sum_k I_2([X_{ijk}]_{i,j}) + \sum_j I_2([X_{ijk}]_{i,k}) \subseteq I_2([X_{ijk}]_{i,(j,k)}).$$

Statisticians have known that $\left(\sum_k I_2([X_{ijk}]_{i,j}) + \sum_j I_2([X_{ijk}]_{i,k}) \right) : (\prod_{i,j,k} X_{ijk})^{\infty} = I_2([X_{ijk}]_{i,(j,k)})$, and they have also known that the latter ideal is a prime ideal not containing any variables; see a proof in Theorem 29. Thus

$$I_2([X_{ijk}]_{i,(j,k)}) = \left(\sum_k I_2([X_{ijk}]_{i,j}) + \sum_j I_2([X_{ijk}]_{i,k}) \right) : (\prod_{i,j,k} X_{ijk})^{\infty},$$

so that the intersection axiom says that one of the associated primes and even primary components of $\sum_k I_2([X_{ijk}]_{i,j}) + \sum_j I_2([X_{ijk}]_{i,k})$ is $I_2([X_{ijk}]_{i,(j,k)})$. Fink in [54] determined all other associated prime ideals of $\sum_k I_2([X_{ijk}]_{i,j}) + \sum_j I_2([X_{ijk}]_{i,k})$, proving the conjecture of Cartwright and Engström (conjecture is stated in [47, p. 146]).

The papers [8] and [111] algebraically generalize the **intersection axiom** to the following: if all for all possible values i_j of Y_j, $P(Y_1 = i_1, \ldots, Y_n = i_n) > 0$, and if $Y_1 \perp\!\!\!\perp Y_i \mid (\{Y_2, \ldots, Y_n\} \setminus \{Y_i\})$ for all $i = 2 \ldots, n$, then $Y_1 \perp\!\!\!\perp \{Y_2, \ldots, Y_n\}$.

3.3   A Version of the Hammersley-Clifford Theorem

For completeness I give in this section the most algebraic proof I can think of of the Hammersley-Clifford Theorem. A different proof can be found in [74, p. 36], and there is more discussion in [47, p. 80].

Let G be an undirected graph on the set of vertices $[n]$. Let $Y_1, \ldots, Y_n$ be discrete random variables. Associated to this graph is a collection of conditional independence statements:

$$\{Y_i \perp\!\!\!\perp Y_j \mid (\{Y_1, \ldots, Y_n\} \setminus \{Y_i, Y_j\}) : i \neq j, (i,j) \text{ is not an edge in } G\}.$$

(Such a graphical model of conditional independence statements is said to satisfy the **pairwise Markov property**.) If Y_i takes on r_i distinct values, then we need $r_1 \cdots r_n$ variables X_a, and we denote by I_G the ideal generated by all the 2×2-minors of all the matrices obtained from all the conditional independence statements (over some understood field F).

For example, if $n = 3$ and the only edge in the graph is $(2, 3)$, the associated conditional independences are

$$Y_1 \perp\!\!\!\perp Y_2 \mid Y_3 \text{ and } Y_1 \perp\!\!\!\perp Y_3 \mid Y_2,$$

which are precisely the hypotheses of the intersection axiom. Fink [54] analyzed the corresponding ideal. Swanson and Taylor [111] analyzed the ideals for arbitrary n and $t \in [n]$ with the graph being the complete graph on vertices $t + 1, \ldots, n$; Ay and Rauh [8] analyzed the case for arbitrary n and $t = 1$.

Theorem 29 (Hammersley and Clifford) *Let n, G, I_G be as above. Then I_G : $(\prod_a X_a)^\infty$ is a binomial prime ideal which does not contain any variables. In particular, I_G : $(\prod_a X_a)^\infty$ is a minimal prime ideal over I_G, and its primary component is the prime ideal.*

Furthermore, the variety of the prime ideal in this theorem has a monomial parametrization, which is explicit in the proof below.

Proof Suppose that Y_i takes on r_i distinct values. Without loss of generality these values are in the set $[r_i]$. If any r_i equals 0 or 1, the conditional independence statements can be rephrased without using that Y_i. So we may assume that all r_i are strictly bigger than 1.

If G is a complete graph on $[n]$, then $I_G = 0$, so that $I_G = 0 = I_G : (\prod_a X_a)^\infty$ is a binomial prime ideal which does not contain any variables. In the sequel we assume that G is not a complete graph, so that I_G is a non-zero (binomial) ideal.

Fix a pair of distinct i, j in $[n]$ such that (i, j) is not an edge in G. Fix $\alpha = (\alpha_1, \ldots, \alpha_n)$, with α_k varying over the possible values of the random variable Y_k. Let M_α be the $r_i \times r_j$ generic matrix whose (k, l)-entry is X_a with $a_i = k, a_j = l$, and all other components in a identical to the corresponding components in α. (Obviously α_i and α_j are not needed to specify M_α.) The ideal I_{ij} expressing the conditional independence statement $Y_i \perp\!\!\!\perp Y_j \mid (\{Y_1, \ldots, Y_n\} \setminus \{Y_i, Y_j\})$ is generated by all $I_2(M_\alpha)$ as α varies.

By definition $I_G = \sum_{i,j} I_{ij}$, as i, j vary over distinct elements of $[n]$ such that (i, j) is not an edge (and without loss of generality $i < j$).

A clique in G is a subset of its vertices any two of which are connected by an edge. For any maximal clique L of G and for each $c_L \in \prod_{i \in L}[r_i]$, let T_{L,c_L} be a variable over the underlying field F. Let $\varphi : F[X_a : a] \to F[T_{L,c_L} : L, c_L]$ be the F-algebra homomorphism such that $\varphi(X_a) = \prod_L T_{L,a(L)}$, as L varies over the maximal cliques of G, and where $a(L)$ is the $|L|$-tuple consisting only of the L-components of a. Let P be the kernel of φ.

Warning: Whereas I_G is the sum of the I_{ij} where (i, j) is not an edge, the variables T_{L,c_L} and thus the map φ instead use (cliques of) edges and isolated vertices.

We prove that $I_G \subseteq P$. It suffices to prove that $I_{ij} \subseteq P$, where (i, j) is not an edge. For simplicity, suppose that $(1, 2)$ is not an edge in G. By reindexing it suffices to prove that $X_{(1,1,\ldots,1)}X_{(2,2,1,\ldots,1)} - X_{(1,2,1,\ldots,1)}X_{(2,1,1,\ldots,1)} \in P$. To simplify notation, we treat below $T_{L,c(L)}$ as 1 if L is not a clique of G. Note that no clique contains both 1 and 2. Then φ maps $X_{(1,1,\ldots,1)}$ to

$$\prod_{1 \in L} T_{L,(1,\ldots,1)} \prod_{2 \in L} T_{L,(1,\ldots,1)} \prod_{1,2 \notin L} T_{L,(1,\ldots,1)},$$

$X_{(2,2,1,\ldots,1)}$ to

$$\prod_{1 \in L} T_{L,(2,1,\ldots,1)} \prod_{2 \in L} T_{L,(2,1,\ldots,1)} \prod_{1,2 \notin L} T_{L,(2,\ldots,1)},$$

$X_{(1,2,1,\ldots,1)}$ to

$$\prod_{1 \in L} T_{L,(1,\ldots,1)} \prod_{2 \in L} T_{L,(2,1,\ldots,1)} \prod_{1,2 \notin L} T_{L,(1,\ldots,1)},$$

and $X_{(2,1,1,\ldots,1)}$ to

$$\prod_{1 \in L} T_{L,(2,1,\ldots,1)} \prod_{2 \in L} T_{L,(1,\ldots,1)} \prod_{1,2 \notin L} T_{L,(2,\ldots,1)},$$

so that $X_{(1,1,\ldots,1)}X_{(2,2,1,\ldots,1)} - X_{(1,2,1,\ldots,1)}X_{(2,1,1,\ldots,1)}$ is mapped by φ to 0. Thus $I_G \subseteq P$.

As φ is a homogeneous monomial map of positive degree, P is generated by binomials and does not contain any variables. It follows that $I_G : (\prod_a X_a)^\infty \subseteq P$.

Now let $f \in P$. The proof below that $f \in I_G : (\prod_a X_a)^\infty$ is fairly elementary, only long in notation. Since P is the kernel of a homogeneous monomial map, we may assume that $f = X_{a_1} \cdots X_{a_m} - X_{b_1} \cdots X_{b_m}$ for some n-tuples $a_1, \ldots, a_m, b_1, \ldots, b_m$. To show that $f \in I_G : (\prod_a X_a)^\infty$, it suffices to prove that any monomial multiple of f is in $I_G : (\prod_a X_a)^\infty$. Fix a non-edge (i, j). Suppose that in a_k neither the ith nor the jth component is 1. Let c_k be the n-tuple whose ith and jth components are 1 and whose other components agree with the components of a_k. Both X_{a_k} and X_{c_k} lie in the same submatrix of $[X_a]_a$ that gives I_{ij}, so that $X_{a_k}X_{c_k}$ reduces modulo I_{ij} and hence modulo I_G to $X_{a'_k}X_{c'_k}$ where a'_k and c'_k each have entry 1 either in the ith or the jth components. Let U be the product of all such X_{c_k}. Then modulo I_G, Thus Uf reduces with respect to I_G to a binomial in which the subscripts of all the variables appearing in the first monomial have at least one of i,j components equal to 1, and in the second monomial the number of non-1ith and jth components in the subscripts does not increase. By repeating this for the second monomial as well, we may assume that for each variable appearing in f, the ith or the jth component in the subscript is 1. If we next similarly clean positions i',j' in this way, we do not at the same time lose the cleaned property of positions i and j: as factors of the multipliers U keep the clean (i,j) property. By repeating this cleaning, in finitely many rounds we get a binomial f in P such that for each non-edge (i,j) and for each variable appearing in f, the ith or the jth component of the subscript of that variable is 1.

With the assumption that for each non-edge (i,j), the ith or the jth component of a_k and of b_k is 1, we claim that $f = 0 \in I_G$. If $a_i = b_j$ for some $i,j \in [m]$, then the binomial f/X_{a_i} has the same property of many components being 1, and it suffices to prove that $f/X_{a_i} = 0 \in I_G$. Thus without loss of generality we may assume that $m > 0$ and that $a_i \neq b_j$ for all $i,j \in [m]$. Let K_j (resp. L_j) be the set of all $i \in [n]$ such that the ith component in a_j (resp. b_j) is not 1. By possibly reindexing we may assume that K_1 is maximal among all such sets. By the assumption on the 1-entries, necessarily K_1 is contained in a maximal clique L of G, and for all $i \in [n] \setminus L$, the ith component in a_1 is 1. Since $f \in P$, the variable $T_{L,a_1(L)}$ must also divide $\varphi(X_{b_k})$ for some $k \in [m]$. This means that a_1 and b_k agree in the L-components, and in particular, $K_1 \subseteq L_k$. By maximality of K_1, necessarily $K_1 = L_k$, whence $a_1 = b_k$, which is a contradiction.

This proves that $P = I_G : (\prod_a X_a)^\infty$ is a binomial prime ideal containing no variables. Thus $I_G : (\prod_a X_a)^\infty$ is contained in the P-primary component of I_G, and since $I_G : (\prod_a X_a)^\infty$ is primary (even prime) and contains I_G, necessarily it is the P-primary component. $\qquad\square$

In particular, if $n = 3$ and the only edge in G is $(2,3)$, then I_G is the ideal of the intersection axiom, which fills in the details in the discussion on page 65. Even more simply, if $n = 2$ and G contains no edges, then $I_G = I_G : (\prod_a X_a)^\infty$ is the ideal generated by the 2×2-minors of the generic matrix.

Remark 30 To any monomial parametrization $\varphi : F[X_c : c] \rightarrow F[T_d : d]$ we can associate a 0–1 matrix A whose (c,d)-entry equals 1 if T_d is a factor of $\varphi(X_c)$, and is 0 otherwise. In the theorem above the indices c were n-tuples; here we assume that these are ordered in some way, so that for any monomial $\prod_c X_c^{e_c}$ we can talk about the exponent vector $(e_c : c)$. For any binomial $\prod_c X_c^{e_c} - \prod_c X_c^{f_c}$ in the kernel of φ, the corresponding vector $(e_c : c) - (f_c : c)$ is in the kernel of A. Conversely, for any integer vector $(e_c : c)$ in the kernel of A, the binomial $\prod_{e_c>0} X_c^{e_c} - \prod_{e_c<0} X_c^{-e_c}$ is a binomial in the kernel of φ. Thus finding a set the kernel of φ is the same as finding the kernel of A as a $\mathbb{Z}$-submodule of the set of all integer vectors. The generating set of the latter kernel is a **Markov basis** for A, and its connections to algebraic statistics were first explored by Diaconis and Sturmfels in [46].

3.4 Summary/Unification of Some Recent Papers

This is a partial summary of the papers Fink [54], Herzog et al. [64], Ohtani [86], Ay-Rauh [8], Swanson and Taylor [111]: there are some similarities in the methods and results of these papers, but there does not seem to be one all-encompassing theorem. I present these results using as much of the common language as I can, but the four papers have further details and results.

Let $r_1, r_2, \ldots, r_n$ be positive integers, and let $N = [r_1] \times [r_2] \times \cdots \times [r_n]$ (where for any positive integer r, $[r] = \{1, 2, \ldots, r\}$). Let R be the polynomial ring in variables X_a over a field, where a varies over elements in N. We will often refer to the generic hypermatrix $[X_a : a \in N]$, so we give it a name, say M.

A generalized two-by-two determinant of M, for given $a, b \in N$ and $K \subseteq [n]$, is

$$f_{K,a,b} = X_a X_b - X_{s(K,a,b)} X_{s(K,b,a)},$$

where $s(K, a, b)$ is an element of N with

$$s(K, a, b)_j = \begin{cases} b_j, & \text{if } j \in K; \\ a_j, & \text{if } j \notin K. \end{cases}$$

If $K = \{i\}$, we also write $s(i, a, b)$ for $s(\{i\}, a, b)$ and $f_{i,a,b}$ for $f_{\{i\},a,b}$. When a and b differ only in positions i and j, then $f_{i,a,b}$ is precisely a standard two-by-two determinant of the submatrix of M obtained by keeping the entries that agree with a and b in the positions $k \neq i, j$.

Let $t \in [n]$. For each $i \in [t]$ let G_i be a simple graph on $[r_1] \times \cdots \times \widehat{[r_i]} \times \cdots \times [r_n]$. (These graphs play a very different role from the ones in Sect. 3.3.) Define

$$I^{(t)}(G_1, \ldots, G_t) =$$

$$(f_{i,a,b} : i \leq t, \{(a_1, \ldots, \widehat{a_i}, \ldots, a_n), (b_1, \ldots, \widehat{b_i}, \ldots, b_n)\} \text{ is an edge in } G_i).$$

These ideals have been studied as follows:

1. Fink [54]: $n = 3, t = 1$, and G_1 is the grid graph on $[r_2] \times [r_3]$, namely $G_1 = \left(\bigcup_{j \in [r_2], k_1, k_2 \in [r_3]} \{(j, k_1), (j, k_2)\} \right) \cup \left(\bigcup_{k \in [r_3], j_1, j_2 \in [r_2]} \{(j_1, k), (j_2, k)\} \right)$.
2. Herzog et al. [64] and independently Ohtani [86]: $n = 2, r_1 = 2, t = 1$.
3. Ay and Rauh [8]: $t = 1$.
4. Swanson and Taylor [111]: for each i, G_i is the grid graph on $[r_1] \times \cdots \times \widehat{[r_i]} \times \cdots \times [r_n]$, i.e., the edges consist of those pairs of $(n-1)$-tuples that differ in precisely one component.

Throughout $t \in [n]$. For each $i \in [t]$, let $N_i = [r_1] \times \cdots \times \widehat{[r_i]} \times \cdots \times [r_n]$, and let G_i be a graph on N_i. We write G for $\{G_1, \ldots, G_t\}$. We use the Hamming distance on N: $d(a,b) = \#\{i \in [n] : a_i \neq b_i\}$, and $D(a,b) = \{i \in [n] : a_i \neq b_i\}$.

Definition 31 We say that $a, b \in N$ are **directly connected relative to G_i** if $\{(a_1, \ldots, \widehat{a_i}, \ldots, a_n), (b_1, \ldots, \widehat{b_i}, \ldots, b_n)\}$ is an edge in G_i.

We say that $a, b \in N$ are **connected relative to G_i** if there exist $c_1, c_2, \ldots, c_{k-1} \in N$ such that with $c_0 = a$ and $c_k = b$, for each $j = 1, \ldots, k$, c_{j-1} and c_j are directly connected relative to G_i. We call $a = c_0, c_1, \ldots, c_{k-1}, c_k = b$ a **path** from a to b relative to G_i.

We say that $a, b \in N$ are **connected relative to G** if there exist $c_1, c_2, \ldots, c_{k-1} \in N$ such that with $c_0 = a$ and $c_k = b$, for each $j = 1, \ldots, k$, there exists $i \in [t]$ such that c_{j-1} and c_j are directly connected relative to G_i. We call $a = c_0, c_1, \ldots, c_{k-1}, c_k = b$ a **path** from a to b relative to G.

Lemma 32 *Let $i \in [t]$ and let $c_0, \ldots, c_k$ be a path relative to G_i. Then*

$$\left(\prod_{j=1}^{k-1} X_{c_j}\right) \cdot f_{i,c_0,c_k} \in I^{\langle t \rangle}(G_i).$$

Proof (Similar Versions of This Are Proved in [8] and [111].) If the ith components in c_0 and c_k are identical then $f_{i,c_0,c_k} = 0$. If c_0, c_k without the ith components form an edge in G_i, then $f_{i,c_0,c_k} \in I^{\langle t \rangle}(G_i)$. In particular, the lemma holds if $k \leq 1$. Now let $k \geq 2$. Then modulo $I^{\langle t \rangle}(G_i)$, with U an abbreviation for $X_{c_1} \cdots X_{c_{k-2}}$,

$$X_{c_0} U X_{c_{k-1}} X_{c_k} \equiv X_{s(i,c_0,c_{k-1})} U X_{s(i,c_{k-1},c_0)} X_{c_k} \text{ (by induction on } k)$$

$$\equiv X_{s(i,c_0,c_{k-1})} U X_{s(i,s(i,c_{k-1},c_0),c_k)} X_{s(i,c_k,s(i,c_{k-1},c_0))}$$

$$\text{(since } s(i, c_{k-1}, c_0), c_k \text{ is a path relative to } G_i)$$

$$= X_{s(i,c_0,c_{k-1})} U X_{s(i,c_{k-1},c_k)} X_{s(i,c_k,c_0)}$$

$$\equiv X_{s(i,s(i,c_0,c_{k-1}),s(i,c_{k-1},c_k))} U X_{s(i,s(i,c_{k-1},c_k),s(i,c_0,c_{k-1}))} X_{s(i,c_k,c_0)}$$

$$\text{(by induction on } k, \text{ since}$$

$$s(i, c_0, c_{k-1}), c_1, \ldots, c_{k-2}, s(i, c_{k-1}, c_k) \text{ is a path relative to } G_i)$$

$$= X_{s(i,c_0,c_k)} U X_{c_{k-1}} X_{s(i,c_k,c_0)},$$

which proves the lemma. □

Remark 33 Note how the ith entry in the path is not important! But if we want to mix G_i and G_j, the ith entries make a difference (and it is not clear how to control for that fully, in fact, the ideals in [111] have embedded primes whose characterization is not complete).

Lemma 34 *Let $i \in [t]$. Let H be the set of all elements of the form $\left(\prod_{j=1}^{k-1} X_{c_j}\right) \cdot f_{i,c_0,c_k}$ as $c_0, \ldots, c_k$ vary over paths relative to G_i. Then H is a (redundant) Gröbner basis in the lexicographic order.*

Proof Let $f = \left(\prod_{j=1}^{k-1} X_{c_j}\right) \cdot f_{i,c_0,c_k}$ and $g = \left(\prod_{j=1}^{l-1} X_{d_j}\right) \cdot f_{i,d_0,d_l}$. We want to show that the S-polynomial of f and g reduces to 0 with respect to H. In the lexicographic order, the leading monomial of f_{i,c_0,c_k} is either $X_{c_0} X_{c_k}$ or $X_{s(i,c_0,c_k)} X_{s(i,c_k,c_0)}$. In the latter case, since $f_{i,c_0,c_k} = -f_{i,s(i,c_0,c_k),s(i,c_k,c_0)}$ and since $s(i,c_0,c_k), c_1, \ldots, c_{k-1}, s(i,c_k,c_0)$ is a path relative to G_i, by possibly replacing c_0 and c_k with their switches we may assume that the leading term of f is $X_{c_0} X_{c_k}$. Similarly we may assume that the leading term of g is $X_{d_0} X_{d_l}$. By standard Gröbner bases, if $\{c_0, c_k\}$ and $\{d_0, d_l\}$ are disjoint, then the S-polynomial of f and g reduces to 0. If $c_0 = d_0$ and $c_k = d_l$, then $S(f,g) = m(X_{s(i,d_0,d_l)} X_{s(i,d_l,d_0)} - X_{s(i,c_0,c_k)} X_{s(i,c_k,c_0)})$, where $m = \mathrm{lcm}(X_{c_1} \cdots X_{c_k}, X_{d_1} \cdots X_{d_l})$ is the product of all the variables in a path from $s(i,d_0,d_l) = s(i,c_0,c_k)$ to $s(i,d_l,d_0) = s(i,c_k,c_0)$, so that this S-polynomial is in H. It remains to consider the case $c_0 = d_0$ and $c_k \neq d_l$. Then $S(f,g) = m(X_{c_k} X_{s(i,d_0,d_l)} X_{s(i,d_l,d_0)} - X_{d_l} X_{s(i,c_0,c_k)} X_{s(i,c_k,c_0)})$, where $m = \mathrm{lcm}(X_{c_1} \cdots X_{c_k}, X_{d_1} \cdots X_{d_l})$. Consider the term $X_{c_k} X_{s(i,d_0,d_l)}$: if it is bigger in the lexicographic order than $X_{s(i,c_k,d_l)} X_{s(i,d_0,c_k)}$, then since m is a product of the right variables in the right path, we can reduce $S(f,g)$ further. Any further reductions of the two degree-three terms in the binomial part can be reduced similarly because m has enough variables, until $S(f,g)$ reduces to 0. $\qquad\square$

Papers [54, 64, 86], and [8] go further and determine minimal Gröbner bases, via further restrictions on admissible paths.

3.5   A Related Game

One would understand the primary components of I_G in the previous section much better if one understood the following:

Problem 35 Let $a_1, \ldots, a_m, b_1, \ldots, b_m$ be n-tuples ($2m$ of them) such that $X_{a_1} \cdots X_{a_m} - X_{b_1} \cdots X_{b_m} \in I^{\langle n \rangle}(G)$. (For ideals in [111], an equivalent and more elementary check for ideal membership is that for each $i = 1, \ldots, n$, the ith components of $a_1, \ldots, a_m$ are up to order the same as the ith components of $b_1, \ldots, b_m$.) Carry out the successive rewriting of $X_{a_1} \cdots X_{a_m}$ with respect to the generators of $I^{\langle n \rangle}(G)$ to get to $X_{b_1} \cdots X_{b_m}$.

Since this is a hard problem, I would like instead somebody to make it a computer game or an app:

Game The computer serves you two lists of n-tuples of positive integers: $a_1, \ldots, a_m$ and $b_1, \ldots, b_m$. (In one version of the game, $X_{a_1} \cdots X_{a_m} - X_{b_1} \cdots X_{b_m} \in I^{\langle n \rangle}(G)$, in another version whether this is so is determined by chance.) The following move is allowed on the list $a_1, \ldots, a_m$: if a_i and a_j differ in exactly two components, say k

and l, replace the list $a_1, \ldots, a_m$ with the list $c_1, \ldots, c_m$ where $c_i = s(k, a_i, a_j) = s(l, a_j, a_i)$, $c_j = s(k, a_j, a_i) = s(l, a_i, a_j)$, and for all $s \neq i, j, c_s = a_s$. Repeat the moves on the new list $c_1, \ldots, c_m$ until you get the list $b_1, \ldots, b_m$. You get bonus points for accomplishing the task in few moves.

I envision users all over the world solving (playing with) instances of this while waiting for a bus or in coffee shops, and they could be competing for the shortest number of moves, with possibly short answers being transmitted to some central station.

3.6 Binomial Edge Ideals with Macaulay2

Let us make now a short review of some of the preceding results with the computer algebra system Macaulay2 at hand.

First, we will make use of the package `Binomials` so we load it into the system:

```
i1 : needsPackage "Binomials"
```

Consider now a simple graph G on n vertices and a polynomial ring in $2n$ variables, for each edge (i, j) we consider the binomial $f(i, j) = x_i y_j - x_j y_i$. The ideal generated by such binomials is the **binomial edge ideal** of G, J_G. We construct it with the following simple Macaulay2 function:

```
i2 : graphminorsedge = (n,LL) -> (
     HHR = QQ[x_1..x_n, y_1..y_n];
     ideal apply(LL, k-> x_(k_0) * y_(k_1) - x_(k_1) * y_(k_0))
)
```

Observe that this is a generalization of the ideal of 2-minors of a $2n$-matrix of indeterminates (which corresponds to the binomial edge ideal of the complete n-graph).

We say that the graph G is **closed** with respect to the labelling if for all (i, j), (k, l) such that $i < j$ and $k < l$ we have another edge (j, l) if $i = k$ and (i, k) if $j = l$. With the help of Macaulay2 the reader can try some examples of ideals of closed graphs and some ideals of non-closed graphs to see how Theorem 1 in [64] works.

A nice exercise is to experiment with closed **bipartite graphs** to find their Gröbner bases.

In general, if the graph is not closed, the Gröbner basis does not coincide with the binomials given by the edges, but can we find the basis in the graph? Let us define admissible paths $i_1, \ldots, i_r$ as follows:

1. $i_k \neq i_l \forall 1 \leq k \neq l \leq r$.
2. For each $k = 1, \ldots, r - 1$ either $i_k < i$ or $i_k > j$.
3. For any proper subset $\{j_1, \ldots, j_s\}$ of $\{i_1, \ldots, i_{r-1}\}$ the sequence $i, j_1, \ldots, j_s, j$ is not a path.

One can write a function to construct all admissible paths and use it to find all the admissible paths in a closed graph. Now, for each admissible path p construct the monomial $u_p = \prod_{(i_k > i)} (x_{i_k}) \prod_{(i_l < i)} (y_{i_l})$. The Gröbner basis is then given by $\bigcup_{i<j} \{u_p f_{i,j} | p \text{ is an admissible path from } i \text{ to } j\}$.

We can follow [86] that describes operations on graphs that lead to a primary decomposition of J_G. First, define complete vertices as those such that all their neighbours are connected among them. We perform the following operations on any vertex v that is not complete:

1. Delete v and all the edges incident to v
2. Add all edges that connect vertices in the neighbourhood of v.

From each of these operations we obtain a graph, G' and G'' respectively, each of one has less non-complete vertices. These graph operations yield algebraic operations:

1. $J_{G'} + (x_v, y_v)$
2. $I_{G''} + I_2(N_G(v))$ where $N_G(v)$ is given by the binomials involving v.

Taking as base case the complete graph, whose ideal is prime, this decomposition leads to an alternative algorithm for primary decompositions. We encourage the reader to use Macaulay2 to write a program that implements Ohtani's procedure and compare the results with the in-built primary decomposition algorithms.

3.7 A Short Excursion Into Networks Using Monomial Primary Decompositions

To finish this chapter, let us enter into the world of networks, bringing primary decompositions with us. We will use primary decompositions of monomial ideals here. The monomial case, simpler than the general polynomial case has however multiple applications. We include this section to add yet another view of the use of primary decompositions. Networks are ubiquitous and there are many different approaches to them. A beautiful survey on the topic is [84]. One can see networks as graphs, where we call vertices to the nodes and edges to the connections. Graphs have been extensively studied using commutative algebra, cf. for example [80, 115]. We will in this section introduce the reader to the use of primary decompositions to study the problem of network resilience, in particular the design of attack strategies to break a connected network into disconnected pieces.

Consider a connected network (graph) N. We want to remove nodes (and all the incident connections) so that the network becomes disconnected as soon as possible. What is a good strategy to choose which nodes to delete first? A simple intuitive strategy is to delete first the nodes with biggest degree (i.e. with most connections incident to it). Other strategies are based on different data like betweenness centrality, etc. The approach we are using in this section is to attack the network based on its vertex covers.

A vertex cover of a graph (we see now networks as graphs) is a set C of vertices such that each edge of the graph is incident to at least one vertex of C. C is a **minimal** vertex cover if no subset of C is a vertex cover. C is a **minimum** vertex cover if it is a vertex cover of minimal cardinality. Minimal and minimum vertex covers are not unique in general. Given a graph G we denote by $mvc(G)$ the set of minimal vertex covers of G and by $MVC(G)$ the set of minimum vertex covers of G. We furthermore denote as $\tau(G)$ cardinality of any minimum vertex cover of G, $\tau(G)$ is called the *covering number* of G. For any vertex v we define the *covering degree* and *covering index* of a vertex n as follows

Definition 36 The **covering degree** of v, denoted $cd(v)$ is the number of minimal vertex covers that contain v,

$$cd(v) := \#\{V' \in mcv(G) \text{ such that } v \in V'\}.$$

The **covering index** of v, denoted $ci(v)$ is computed as the number of minimum vertex covers that contain v plus the ratio of the number of minimal vertex covers that contain v to the total number of minimal vertex covers of G,

$$ci(v) := \#\{V' \in MCV(G) \text{ such that } v \in V'\} + \frac{cd(v)}{|mcv(G)|}$$

Two strategies to break up our graph (network) G consist in deleting first the node with highest covering degree or to delete first the vertex with highest covering index, and then proceed downwards. These strategies have been proven to be efficient in several network models [100].

To use these strategies we need to compute all minimal and/or minimum vertex covers, which is a difficult problem in general (it is an example of an NP-hard problem). Here is were computational commutative algebra can help. To every graph G one can associate its edge ideal I_G [115], which is a monomial ideal. Every primary component (equivalently every generator of its Alexander dual) corresponds to a minimal vertex cover of G. One can see that the covering number of G is exactly the codimension of I_G. With these correspondences at hand one can then use a computer algebra system to compute covering degree and covering index of every vertex of G and employ the described strategies.

Example 37 Let G be a line graph with three nodes x, y, z and two edges $(x, y), (y, z)$. It is clear that G has four vertex covers $\{x, y\}, \{x, z\}, \{y, z\}$ and $\{y\}$ but only two of them are minimal, $\{x, z\}$ and $\{y\}$, and only the last one is a minimum vertex cover.

We will use the Macaulay2 package `EdgeIdeals` and compute the algebraic equivalent to the above description:

```
i1 : loadPackage "EdgeIdeals";

i2 : R=QQ[x,y,z];

i3 : G=graph {{x,y},{y,z}};

i4 : I=edgeIdeal G;
```

```
i5 : primaryDecomposition I

o5 = {monomialIdeal(y), monomialIdeal (x, z)}

i6 : codim I

o6 = 1

i7 : dual(I)

o7 = monomialIdeal (y, x*z)
```

As the number of vertices in the graph grows, the number of minimal vertex coverings grows exponentially, to say it algebraically, as the number n of variables grows, the number of primary components a monomial ideal in n variables (i.e. the number of generators of it Alexander dual) grows exponentially. A known (achievable) higher bound is $3^{\frac{n}{3}}$. So computing covering degree and index is expensive in general. Is there any advantage in using the strategies based on covering index and degree instead of using just vertex degree for example? As example 37 shows, it might happen that vertex degree and covering degree or index are correlated and the result of using vertex degree is similar, while the computational effort is much smaller. Experiments show, however, see [100] that vertex degree or betweenness centrality are not correlated to covering degree and index in several types of network models. Furthermore, the attacks based on covering degree and index are far more efficient that those based on vertex degree or betweenness centrality.

We propose the reader to experiment with the primary decompositions of edge ideals of some structured graphs, like the wheel n-graph or with random network models, such as Erdös-Renyi, Watts-Strogatz or Albert-Barabasi.

Combinatorics and Algebra of Geometric Subdivision Operations

Fatemeh Mohammadi and Volkmar Welker

1 Introduction

In the subsequent sections we survey results from combinatorics, discrete geometry and commutative algebra concerning invariants and properties of subdivisions of simplicial complexes. For most of the time we are interested in deriving results that hold for specific subdivision operations that are motivated from combinatorics, geometry and algebra. In particular, we study barycentric, edgewise and interval subdivisions (see Sect. 3 for the respective definitions). Even though we mention some suspicion that part of the results we present may only be a glimpse of what is true for general subdivision operations we do not focus on this aspect. In particular, we are quite sure that some asymptotic results and some convergence results from Sect. 9 are just instances of more general phenomena. Overall, retriangulations are subtle geometric operations and we refer the reader to the book [40] for a comprehensive introduction. Since our focus lies on specific constructions we make only little use of the theory from [40]. Nevertheless, we are convinced that if one wants to go beyond specific subdivision operations it will become inevitable to dig deeper into the theory of triangulations.

We start in Sect. 2 with a quick introduction on abstract and geometric simplicial complexes. For most of the paper we work with abstract simplicial complexes but for some definitions and perspectives the geometric viewpoint turns out to be advantageous. In Sect. 3 we introduce the concept of a subdivision and the three

F. Mohammadi
School of Mathematics, University of Bristol, Bristol BS8 1TW, UK
e-mail: fatemeh.mohammadi@bristol.ac.uk

V. Welker (✉)
Fachbereich Mathematik und Informatik, Philipps-Universität Marburg, Hans-Meerwein-Strasse, 35032 Marburg, Germany
e-mail: welker@mathematik.uni-marburg.de

© Springer International Publishing AG 2017
A.M. Bigatti et al. (eds.), *Computations and Combinatorics in Commutative Algebra*, Lecture Notes in Mathematics 2176, DOI 10.1007/978-3-319-51319-5_3

guiding examples which are considered in our text. We define barycentric, edgewise and interval subdivision, the latter being a special case of a subdivision operation studied in differential geometry. In Sect. 4 we introduce the algebraic side of the picture. This side centers around the Stanley-Reisner ring of a simplicial complex Δ. We also introduce the basic enumerative invariants of a simplicial complex relevant for this manuscript—the f- and the h-vector of a simplicial complex and their relation to the Hilbert-series of the Stanley-Reisner ring. With this preparation in Sect. 5 we can provide the known results on the effect of three subdivision operations on the f- and h-vector of a simplicial complex. The following Sect. 6 lists combinatorial and algebraic invariants and properties of simplicial complexes, and describes when they are invariant under subdivisions. Then in Sect. 7 properties of the h-vector that arise after a few subdivisions are studied. This is shown to relate to algebraic properties of Veronese algebras and the analytic behavior of the h-polynomial. In particular, polynomials with real roots are in the spotlight: we explain how they are tied to Koszul algebras and the Charney-Davis conjecture. In Sect. 8 we approach the behavior of f- and h-vectors after a few subdivisions from the point of view of Lefschetz properties of quotients of the Stanley-Reisner ring by a regular sequence of linear forms. Besides exhibiting results we speculate about connections of consequences of the Lefschetz property and real rootedness. Then in Sect. 9 we study the behavior of h- and f- vectors when the number of subdivisions goes to infinity. In addition we present results on the limiting behavior of graded Betti numbers of the Stanley-Reisner ring under subdivisions. Finally, in Sect. 10 we study how subdivisions can be used to define free resolutions of monomial ideals. We show that in this context arrangements of hyperplanes appear as a natural object that induce subdivisions which support resolutions. Therefore, the section also contains an introduction to cellular resolutions and some basics about arrangements of hyperplanes.

We complement our text by a list of problems, whose difficulty reaches from simple to serious research level. We add some Macaulay2 [62] sessions whenever explicit computations are feasible. We assume little background knowledge and refer the reader to the survey article [19] for f- and h-vector theory of simplicial complexes, to [29] and [91] for background on commutative algebra and to [81] for background in algebraic topology.

We do not cover Stanley's theory of local h-vectors. This is an important theory and may relate to many aspects of subdivisions we discuss here. There is an excellent recent survey of old and recent developments in this field by Athanasiadis and we refer the reader to [7] and [6]. Also there are interesting non-simplicial subdivision operations. In particular, cubical subdivision operations appear to be well structured and interesting objects. First results in the spirit of this survey can be found in [101].

2 Abstract and Geometric Simplicial Complexes

An abstract simplicial complex Δ over the ground set Ω is a subset $\Delta \subseteq 2^{\Omega}$ of the powerset of Ω such that $A \subseteq B \in \Delta$ implies $A \in \Delta$. All simplicial complexes that are of interest in this text are over finite ground set Ω and therefore from now on we will always implicitly assume that a simplicial complex is over a finite ground set and hence finite itself. The elements $F \in \Delta$ are called the faces of Δ and the inclusionwise maximal faces are called facets. The dimension of a face F is $\dim F := \#F - 1$ and the dimension $\dim \Delta$ of Δ is $\max_{F \in \Delta} \dim F$.

Besides this combinatorial aspect of simplicial complexes there is also a geometric aspect. For this recall that a geometric $(d-1)$-dimensional simplex in $\mathbb{R}^n$ is the convex hull

$$\operatorname{conv}\{v_1, \ldots, v_d\} := \left\{ \sum_{i=1}^{d} \lambda_i v_i : \begin{matrix} \lambda_1, \ldots, \lambda_d \geq 0 \\ \lambda_1 + \cdots + \lambda_d = 1 \end{matrix} \right\}$$

of d affinely independent vectors $v_1, \ldots, v_d$. A face of $\operatorname{conv}\{v_1, \ldots, v_d\}$ is the convex hull of a subset of $\{v_1, \ldots, v_d\}$. In particular, any face of $\operatorname{conv}\{v_1, \ldots, v_d\}$ is again a geometric simplex. Here we consider the empty set as the convex hull of the empty set and the empty set as a face of a geometric simplex. The 0-dimensional vertices are the singletons $\{v_i\}$ for $1 \leq i \leq d$ and the v_i are called the vertices of the geometric simplex. A geometric simplicial complex Γ is a collection of geometric simplices in some $\mathbb{R}^d$ such that

1. if $\sigma \in \Gamma$ and τ is a face of σ then $\tau \in \Gamma$.
2. if $\sigma, \tau \in \Gamma$ then $\sigma \cap \tau$ is a face of both σ and τ.

Analogous to the case of abstract simplicial complexes, we call the elements of a geometric simplicial complex Γ the faces of Γ. The vertex set of a geometric simplicial complex Γ is the collection of all vertices of faces of Γ.

The vertex scheme $\Delta(\Gamma)$ of Γ is the collection of all vertex sets of simplices $\sigma \in \Gamma$. It is immediate from the above definitions that $\Delta(\Gamma)$ is a simplicial complex. If Δ is an abstract simplicial complex and Γ a geometric simplicial complex such that after a suitable relabeling of the vertices we have that $\Delta(\Gamma) = \Delta$ then we say that Γ is a geometric realization of Δ. We consider the union $\bigcup_{\sigma \in \Gamma} \sigma \subseteq \mathbb{R}^d$ as a topological space with the subspace topology inherited from the Euclidean topology on $\mathbb{R}^d$. It is a well known basic fact from topology that every simplicial complex has a geometric realization and that any two geometric realizations are homeomorphic. Therefore, it is unambiguous to write $|\Delta|$ to denote any geometric realization of Δ. In the sequel we will write Δ_{d-1} for an abstract $(d-1)$-simplex, i.e., the power set of a d-element set, and Γ_{d-1} for the standard geometric $(d-1)$-simplex, i.e., the convex hull $\operatorname{conv}(e_1, \ldots, e_d)$ of the d unit vectors $e_1, \ldots, e_d$ in $\mathbb{R}^d$.

Given two simplicial complexes $\Delta(1)$ and $\Delta(2)$ such that their geometric realizations $|\Delta(1)|$ and $|\Delta(2)|$ are homeomorphic, the relation between the combinatorial and the algebraic invariants of $\Delta(1)$ and $\Delta(2)$ is subtle and complicated. We will be interested in the situation when $\Delta(1)$ is a *refinement* of $\Delta(2)$. Given two geometric simplicial complexes $\Gamma(1), \Gamma(2)$ in $\mathbb{R}^d$, we say that $\Gamma(1)$ is a subdivision of $\Gamma(2)$ if $\bigcup_{\sigma \in \Gamma(1)} \sigma = \bigcup_{\sigma \in \Gamma(2)} \sigma$ and every simplex $\sigma \in \Gamma(2)$ is a union of simplices in $\Gamma(1)$. Now we say that an abstract simplicial complex $\Delta(1)$ is a subdivision of the abstract simplicial complex $\Delta(2)$ if there are geometric realizations $\Gamma(1)$ of $\Delta(1)$ and $\Gamma(2)$ of $\Delta(2)$ such that $\Gamma(1)$ is a subdivision of $\Gamma(2)$. Note that even though our definition of subdivision is the most common definition in the topology literature, there are more general concepts of subdivision (see e.g. [105]).

3   Subdivisions of Simplicial Complexes

In this section we list a few well known subdivision operations on simplicial complexes. Clearly, this list is not exhaustive and for sure there are many more such operations lurking in the literature. Rather, we concentrate on three subdivision operations. Two of them have been shown to exhibit particularly nice properties in our context and the third is still mostly unexplored (Fig. 1).

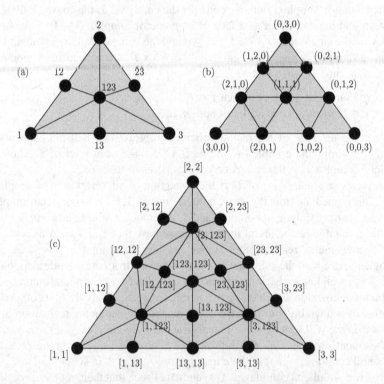

Fig. 1 Barycentric (**a**), 3rd edgewise (**b**), interval (**c**) subdivision of a 2-simplex

3.1   Barycentric Subdivision

The barycentric subdivision of a geometric simplicial complex can be described as follows. Let $v_1, \ldots, v_d \in \mathbb{R}^n$ be affinely independent. For $\emptyset \neq A \subseteq \{v_1, \ldots, v_d\}$ let

$$ b_A = \frac{1}{\#A} \sum_{v \in A} v $$

be the barycenter of the simplex conv(A). Then for any chain $\emptyset \subset A_1 \subset \cdots \subset A_l$ of subsets of $\{v_1, \ldots, v_d\}$ let $\sigma_{A_1 \subset \cdots \subset A_l}$ be the convex hull conv($b_{A_1}, \ldots, b_{A_l}$). For a geometric simplicial complex Γ with vertex scheme $\Delta(\Gamma)$ the set of simplices $\sigma_{A_1 \subset \cdots \subset A_l}$ for chains of subsets $\emptyset \subset A_1 \subset \cdots \subset A_l$ from $\Delta(\Gamma)$ defines a subdivision of Γ which is called the barycentric subdivision of Γ. We write sd(Γ) for the barycentric subdivision of Γ. If Δ is an abstract simplicial complex then define its barycentric subdivision as the simplicial complex sd(Δ) over the ground set $\Delta \setminus \{\emptyset\}$ whose simplices are the subsets $\{A_1, \ldots, A_l\}$ of $\Delta \setminus \{\emptyset\}$ for which with a suitable numbering $A_1 \subset \cdots \subset A_l$. It is easy to verify that $V(\mathrm{sd}(\Gamma))$, the vertex scheme of the barycentric subdivision of a geometric simplicial complex Γ, is (up to relabelling the vertices) the barycentric subdivision of the vertex scheme of Γ, sd($\Delta(\Gamma)$). Barycentric subdivision is a classical subdivision operation from topology. Some of its many applications can be found in texts on algebraic topology such as [81].

3.2   Edgewise Subdivision

Barycentric subdivision is easily described but has some geometric flaws. In particular, the volumes of the $(d-1)$-simplices in a barycentrically subdivided geometric $(d-1)$-simplex differ. A subdivision that does not have this problem is the edgewise subdivision. It is best explained for geometric $(d-1)$-simplices. The general case then follows after one has checked that it is possible to patch the subdivided simplices. Edgewise subdivisions exist for all natural numbers $r \geq 1$. Let $r \geq 1$ then the rth edgewise subdivision of the $(d-1)$-simplex Γ_{d-1} is defined as follows. Consider the rth dilation $r\Gamma_{d-1}$ of the $(d-1)$ simplex with vertices the unit basis vectors in $\mathbb{R}^d$. The integer points in $r\Delta_{d-1}$ are the d-tuples $(i_1, \ldots, i_d)$ of non-negative integers such that $i_1 + \cdots + i_d = r$. We write $\Omega_{d,r}$ for this set. Now we make a change of coordinates and map $(i_1, \ldots, i_d)$ to $\iota(i_1, \ldots, i_d) = (i_1, i_1 + i_2, \ldots, i_1 + \cdots + i_d)$. We subdivide $r\Gamma_{d-1}$ by simplices conv(A) where $A \subseteq \Omega_{d,r}$ and either $\iota(v - v') \in \{0,1\}^d$ or $-\iota(v - v') \in \{0,1\}^d$ for all $v, v' \in A$. Now the rth edgewise subdivision of Γ_{d-1} is obtained from this subdivision of $r\Gamma_{d-1}$ by dilating with factor $\frac{1}{r}$. In general, the rth edgewise subdivision $\Gamma^{(r)}$ is obtained from Γ by applying it to every simplex in Γ. The geometric rth edgewise subdivision clearly induces a subdivision on the vertex scheme of Γ. This way we can speak of the rth edgewise subdivision $\Delta^{\langle r \rangle}$ of an abstract simplicial complex Δ. The term rth edgewise subdivision is motivated by the fact that edges of Γ are subdivided into r equal pieces in $\Gamma^{\langle r \rangle}$. Edgewise subdivision first appeared in a paper by Freudenthal [55] but has found numerous applications in discrete geometry [48], K-theory [61] or commutative algebra [27]. We explain the latter in more detail later in the text.

3.3 Interval Subdivision

This subdivision operation is easiest described starting with an abstract simplicial complex Δ. First, we consider $\Delta \setminus \{\emptyset\}$ as a partially ordered set ordered by inclusion. Let Ω be the set of formal symbols $[A, B]$ for any inclusion $A \subseteq B$ in $\Delta \setminus \{\emptyset\}$. Note $A = B$ is permitted. Now we consider the partial order on the intervals induced by containment and define $\text{Int}(\Delta)$ to be the simplicial complex of all chains of intervals in this order. By Walker [117, Theorem 6.1. (a)] we obtain that $\text{Int}(\Delta)$ is a subdivision of Δ. Indeed this subdivision also appears as the special case $N = 1$ of the subdivision from [32, Fig. 1.2].

Problem 1 Prove that $\text{Int}(\Delta)$ coincides with the subdivision from [33, Fig. 1.2] if one sets $N = 1$ in [33].

4   The Stanley-Reisner Ring

The algebraic object usually associated to a simplicial complex Δ is the face or Stanley-Reisner ring $\mathbb{K}[\Delta]$ of Δ. Let Δ be a simplicial complex over ground set Ω. The ring $\mathbb{K}[\Delta]$ is the quotient of the polynomial ring $\mathbb{K}[x_\omega : \omega \in \Omega]$ over the field $\mathbb{K}$ and the Stanley-Reisner ideal I_Δ generated by the $\mathbf{x}_N := \prod_{i \in N} x_i$ for minimal non-faces N of Δ. Note that a subset $N \subseteq \Omega$ is a minimal non-face of Δ if $N \notin \Delta$ and all proper subsets of N are in Δ.

Since we will have to deal with monomials and ideals generated by monomials more often in this text, we introduce some notation here. A monomial in $\mathbb{K}[x_\omega : \omega \in \Omega]$ is a product $\prod_{\omega \in \Omega} x_\omega^{\alpha_\omega}$ for some non-negative integers α_ω. We also write $\mathbf{x}^\alpha$ for $\prod_{\omega \in \Omega} x_\omega^{\alpha_\omega}$ where $\alpha = (\alpha_\omega)_{\omega \in \Omega}$. In this notation we have for the squarefree monomials $\mathbf{x}_N$ introduced above the identity $\mathbf{x}_N = \mathbf{x}^\alpha$ for $\alpha = \sum_{\omega \in N} e_\omega$ for the unit basis vectors $(e_\omega)_{\omega \in \Omega}$ of $\mathbb{R}^\Omega$. The support $\text{supp}(\mathbf{x}^\alpha)$ of a monomial $\mathbf{x}^\alpha$ is the set $\{\omega : \alpha_\omega \neq 0\}$. An ideal I in $\mathbb{K}[x_\omega : \omega \in \Omega]$ is called a monomial ideal if it is generated by monomials. The Stanley-Reisner ideals are exactly the monomial ideals generated by squarefree monomials $\mathbf{x}_N$ for some collection of subsets $N \subseteq \Omega$.

The first combinatorial invariant we are interested in is the f-vector of Δ. We set $f_i^{\Delta} = \#\{F \in \Delta : \dim F = i\}$ for $-1 \leq i \leq \dim \Delta$ and call $\mathfrak{f}^{\Delta} = (f_{-1}^{\Delta}, \ldots, f_{\dim \Delta}^{\Delta})$ the f-vector of Δ. Sometimes it turns out to be advantageous to encode the f-vector in the f-polynomial $\mathfrak{f}^{\Delta}(t) = \sum_{i=0}^{\dim \Delta + 1} f_{i-1}^{\Delta} t^{\dim \Delta + 1 - i}$. On the algebraic side the Hilbert-series of $\mathbb{K}[\Delta]$ encodes the same information.

For later use, we introduce the Hilbert-series in the generality of standard graded algebras A. A standard graded $\mathbb{K}$-algebra A is a $\mathbb{K}$-algebra that admits a decomposition $A = \bigoplus_{i \geq 0} A_i$ as vector spaces such that $A_0 = \mathbb{K}$, $\dim_{\mathbb{K}} A_1 < \infty$, $A_i A_j \subseteq A_{i+j}$ for all i, j and A is generated by A_1 as a $\mathbb{K}$-algebra. Then the Hilbert-series of A is the formal power series

$$\mathrm{Hilb}_A(t) = \sum_{i \geq 0} \dim_{\mathbb{K}} A_i t^i.$$

By the result of Hilbert and Serre $\mathrm{Hilb}_A(t) = \frac{h_A(t)}{(1-t)^d}$ for some polynomial $h_A(t)$ where $h_A(1) \neq 0$ and d is the Krull dimension $\dim A$ of A. The polynomial $h_A(t)$ is sometimes called the h-polynomial of A.

Any Stanley-Reisner ring $A = \mathbb{K}[\Delta]$ is standard graded where $A = \bigoplus_{i \geq 0} A_i$ with A_i the $\mathbb{K}$-vectorspace spanned by the images of the monomials of degree i in $\mathbb{K}[\Delta]$. Given this description all conditions follow immediately. The vectorspace A_i has a basis formed by the images of the monomials $\underline{x}^{\alpha}$ with support contained in Δ. From this it follows that

$$\dim A_i = \sum_{j=1}^{d-1} f_{j-1}^{\Delta} \binom{i-j+j-1}{j-1}$$

and by a simple calculation that

$$\mathrm{Hilb}_{\mathbb{K}[\Delta]}(t) = \frac{t^d \mathfrak{h}^{\Delta}(1/t)}{(1-t)^d}$$

where $\mathfrak{h}^{\Delta}(t) = \mathfrak{f}^{\Delta}(t-1)$. The polynomial $\mathfrak{h}^{\Delta}(t) = h_0^{\Delta} t^d + \cdots + h_d^{\Delta}$ is called the h-polynomial of Δ and its coefficient vector $\mathfrak{h}^{\Delta} = (h_0^{\Delta}, \ldots, h_d^{\Delta})$ is called the h-vector of Δ. Since $\mathfrak{h}^{\Delta}(1) = \mathfrak{f}^{\Delta}(0) = f_{d-1}^{\Delta} > 0$ it follows that the h-polynomial $h_{\mathbb{K}[\Delta]}(t)$ of $\mathbb{K}[\Delta]$ is the reciprocal polynomial $t^d \mathfrak{h}^{\Delta}(1/t)$ of $\mathfrak{h}^{\Delta}(t)$. The degree of the h-polynomial is always d for a $(d-1)$-dimensional simplicial complex Δ but the h-polynomial of $\mathbb{K}[\Delta]$ has degree $\leq d$. Moreover, it follows that the Krull dimension $\dim \mathbb{K}[\Delta]$ equals $\dim \Delta + 1$.

Example 2 In the following example we compute $\mathrm{Hilb}_{\mathbb{K}[\mathrm{sd}(\Delta_2)]}(t)$, the Hilbert-series of the Stanley-Reisner ring of a barycentrically subdivided 2-simplex.

```
i1 : r:=QQ[x1,x2,x3,x12,x13,x23,x123];

i2 : i:=ideal(x1*x2,x1*x3,x2*x3,x1*x23,x2*x13,x3*x12,x12*x13,
            x12*x23,x13*x23);

o2 : Ideal of QQ[x1, x2, x3, x12, x13, x23, x123]

i3 : H:=hilbertSeries(r/i)

          2      3      4      6
      1 - 9T  + 16T  - 9T  + T
o3 = --------------------------
                  7
            (1 - T)

o3 : Expression of class Divide

i4 : reduceHilbert(H)

               2
      1 + 4T + T
o4 = -----------
               3
        (1 - T)         •
```

Here we calculated the generators of the Stanley-Reisner ideal by hand. Most often in combinatorics, simplicial complexes are given in terms of facets. The conversion of the facet description of a simplicial complex to minimal non-faces description or equivalently its Stanley-Reisner ideal is algorithmically a non-trivial task. Therefore, in most cases it is more convenient to use the Macaulay2 package "SimplicialComplexes" that allows to extract algebraic and enumerative information.

```
i1 : loadPackage("SimplicialComplexes");

i2 : r:=QQ[x1,x2,x3,x12,x13,x23,x123];

o2 : PolynomialRing

i3 : SD:={x1*x12*x123,x1*x13*x123,x2*x12*x123,x2*x23*x123,
          x3*x13*x123,x3*x23*x123};

o3 : List

i4 : SubDiv:=simplicialComplex SD

o4 = | x3x23x123 x2x23x123 x3x13x123 x1x13x123 x2x12x123
       x1x12x123 |

o4 : SimplicialComplex

i5 : fVector SubDiv
```

```
o5 = HashTable{-1 => 1}
                0 => 7
                1 => 12
                2 => 6

o5 : HashTable

i6 : i:=monomialIdeal SubDiv

o6 = monomialIdeal (x1*x2, x1*x3, x2*x3, x3*x12, x2*x13,
                     x12*x13, x1*x23, x12*x23, x13*x23)

o6 : MonomialIdeal of QQ[x1, x2, x3, x12, x13, x23, x123]
```

From here one can proceed to calculate the Hilbert-series as above.

5 The f- and h-Vector Transformations

In this section we study in a purely combinatorial way for a subdivision operator Sub the transformations $F_{\text{Sub}} : \mathfrak{f}^{\Delta} \mapsto \mathfrak{f}^{\text{Sub}(\Delta)}$ and $H_{\text{Sub}} : \mathfrak{h}^{\Delta} \mapsto \mathfrak{h}^{\text{Sub}(\Delta)}$.

5.1 The f-Vector Transformation

Let us first consider the f-vector transformation which is usually much easier to analyze.

5.1.1 Barycentric Subdivision

In order to count the i-dimensional faces of $\text{sd}(\Delta)$ we need to count the number of chains $\emptyset \neq A_1 \subset \cdots \subset A_{i+1}$ of simplices $A_i \in \Delta$. Let us fix a j-dimensional simplex A. We identify a chain $\emptyset \neq A_1 \subset \cdots \subset A_{i+1} = A$ with the ordered set partition $A_1 | A_2 \setminus A_1 | \cdots | A_{i+1} \setminus A_i$ of A. This clearly gives a bijection between i-dimensional faces of $\text{sd}(\Delta)$ and pairs $(A, B_1 | \cdots | B_{i+1})$ where A is a j-dimensional face of Δ and $B_1 | \cdots | B_{i+1}$ an ordered set partition of A into $i + 1$ blocks. We denote by $S(n, \ell)$ the Stirling number of the second kind which counts the number of (unordered) set partitions of an n-element set in ℓ blocks.

Lemma 3 Let Δ be a $(d-1)$-dimensional simplicial complex. Then for $-1 \leq i \leq d-1$

$$\mathfrak{f}_i^{\text{sd}(\Delta)} = \sum_{j=i}^{d-1} (i+1)! S(j+1, i+1) \mathfrak{f}_j^{\Delta}.$$

Note that here we use the convention $S(0, i+1) = 0$ for $i+1 \neq 0$ and $S(0,0) = 1$, which yields $f_{-1}^{\mathrm{sd}(\Delta)} = f_{-1}^{\Delta}$.

5.1.2 Edgewise Subdivision

Here the argumentation already becomes more involved. A very nice argument leading to a formula for the number of i-dimensional simplices in the rth edgewise subdivision can be found in [48, Simplex Counting Lem.]. Even though their final formula is concise and pleasing it is presented in the form of an alternating sum. Therefore, we present the result in the form presented in [26, Proposition 4.2].

Lemma 4 *Let Δ be a $(d-1)$-dimensional simplicial complex and $r \geq 1$. Then, for $-1 \leq i \leq d-1$,*

$$f_i^{\Delta^{\langle r \rangle}} = \sum_{j=i}^{d-1} \sum_{\substack{\ell_1 + \cdots + \ell_{i+1} = j+1 \\ \ell_1 + \cdots + \ell_{i+1} \geq 1}} \binom{r-1}{\ell_1 - 1}\binom{r}{\ell_2} \cdots \binom{r}{\ell_{j+1}} f_j^{\Delta},$$

5.1.3 Interval Subdivision

In order to obtain an i simplex of $\mathrm{Int}(\Delta)$ we need a chain $[A_1, B_1] \subset \cdots \subset [A_{i+1}, B_{i+1}]$ in $\mathrm{Int}(\Delta)$. This translates into a chain $A_{i+1} \subseteq \cdots \subseteq A_1 \subseteq B_1 \subseteq \cdots \subseteq B_{i+1}$ of simplices in Δ such that for all $1 \leq \ell \leq i$ at least one of the inclusions $A_{\ell+1} \subseteq A_\ell$ and $B_\ell \subseteq B_{\ell+1}$ is proper. If we fix the dimension $j = \dim B_{i+1}$ of B_{i+1} then the number of such chains in Δ only depends on f_j^{Δ}. Thus we get a formula:

$$f_i^{\mathrm{Int}(\Delta)} = \sum_{j=-1}^{d-1} c_{i,j} f_j^{\Delta}$$

for some non-negative numbers $c_{i,j}$.

Example 5 Consider the case of 2-dimensional simplicial complexes.

- A 0-simplex $F \in \Delta$ will contribute a single 0-simplex $[F, F]$ in $\mathrm{Int}(\Delta)$.
- A 1-simplex $F = \{a, b\} \in \Delta$ will contribute the three 0-simplices $[F, F]$, $[\{a\}, F]$, $[\{b\}, F]$ and four 1-simplices $[\{a\}, \{a\}] \subset [\{a\}, F]$, $[\{b\}, \{b\}] \subset [\{b\}, F]$ and $[F, F] \subset [\{a\}, F]$, $[F, F] \subset [\{b\}, F]$ in $\mathrm{Int}(\Delta)$.
- A 2-simplex $F = \{a, b, c\}$ will contribute the seven 0-simplices $[A, F]$ for the 7 non-empty subsets of F. For each of the six 0 or 11-dimensional faces A contained in F we have two 1-simplices $[A, A] \subset [A, F]$, $[F, F] \subset [A, F]$, for each of the six chains $A \subset B \subset F$ of a 0-simplex A and a 1-simplex B we have the three 2-simplices $[A, B] \subset [A, F]$, $[B, F] \subseteq [A, F]$, $[B, B] \subseteq [A, F]$ in $\mathrm{Int}(\Delta)$.

This yields 301-simplices per 2-simplex. Now a little computation shows that there will be 242-simplices per 2-simplex in Δ.

This yields the following transformation:

$$f_{-1}^{\mathrm{Int}(\Delta)} = f_{-1}^{\Delta}, \tag{1}$$

$$f_0^{\mathrm{Int}(\Delta)} = f_0^{\Delta} + 3f_1^{\Delta} + 7f_2^{\Delta}, \tag{2}$$

$$f_1^{\mathrm{Int}(\Delta)} = 4f_1^{\Delta} + 30f_2^{\Delta}, \tag{3}$$

$$f_2^{\mathrm{Int}(\Delta)} = 24f_2^{\Delta}. \tag{4}$$

In matrix form this reads:

$$\begin{pmatrix} 1 & 0 & 0 & 0 \\ 0 & 1 & 3 & 7 \\ 0 & 0 & 4 & 30 \\ 0 & 0 & 0 & 24 \end{pmatrix}.$$

Problem 6 Provide a 'nice' formula for $c_{i,j}$.

The same question arises for the case of general N for the subdivision from [32, Fig. 1.2], where the $c_{i,j}$ are replaced polynomials in N whose evaluation at $N = 1$ is $c_{i,j}$.

Let Sub be one of the three subdivision operations for which we have studied the f-vector transformation. In either case the transformation $F_{\mathrm{Sub}} : \mathfrak{f}^{\Delta} \mapsto \mathfrak{f}^{\mathrm{Sub}(\Delta)}$ is a linear transformation. Since in either case $f_i^{\mathrm{Sub}(\Delta)}$ is expressed as a non-negative linear combination of f_j^{Δ} for $j \geq i$ it follows that the matrix representing F_{Sub} is upper triangular with non-negative entries, when choosing for a vector in $\mathbb{R}^{d+1}$ coordinates $(f_{-1}, \ldots, f_{d-1})$. In addition, $f_{-1}^{\Delta} = f_{-1}^{\mathrm{Sub}(\Delta)} = 1$ and for $i \geq 0$ the coefficients of f_i^{Δ} in the expansion of $f_i^{\mathrm{Sub}(\Delta)}$ are pairwise different. The latter is obvious from the Lemmas 3 and 4 for barycentric and edgewise subdivision. For interval subdivision it has to be checked. These facts about the transformation $F_{\mathrm{Sub}} : \mathfrak{f}^{\Delta} \mapsto \mathfrak{f}^{\mathrm{Sub}(\Delta)}$ imply that it is an invertible diagonalizable linear transformation with positive integer eigenvalues where the eigenvalue 1 appears with multiplicity 2 and all other eigenvalues with multiplicity 1.

5.2 The h-Vector Transformation

Now we turn to the h-vector transformation. While the analysis of the f-vector transformation was based on simple counting arguments the proofs of the results on h-vector transformations already require more advanced combinatorial techniques.

5.2.1 Barycentric Subdivision

For the formulation of the result on the transformation we need some notation from permutation statistics. For a permutation $\pi \in S_n$ we say that $i \in [n-1]$ is a descent if $\pi(i) > \pi(i+1)$. For $n \geq 1$ and we define $A(n, j, i)$ as the number of permutations $\sigma \in S_n$ with j descents and $\sigma(1) = i$. We refer the reader to [108, Chap. 1.6] for more details on descents and the Eulerian numbers $A_{n,j} = \sum_{i=0}^{d} A(n, j, i)$.

Theorem 7 ([25, Theorem 1]) *Let Δ be a $(d-1)$-dimensional simplicial complex. Then, for $0 \leq i \leq d$,*

$$h_i^{\mathrm{sd}(\Delta)} = \sum_{j=0}^{d} A(d+1, j, i+1) h_i^{\Delta}.$$

5.2.2 Edgewise Subdivision

For the result in the case of edgewise subdivision we need the following notation:

$$C(r, d, i) = \#\left\{ (a_1, \ldots, a_d) \in \mathbb{N}^d \mid \begin{array}{l} a_1 + \cdots + a_d = i \\ a_j \leq r \text{ for } 1 \leq j \leq d \end{array} \right\}.$$

Theorem 8 ([26, Theorem 1.1]) *Let Δ be a $(d-1)$-dimensional simplicial complex and $r \geq 1$. Then,*

$$h_i^{\Delta^{(r)}} = \sum_{j=0}^{d} C(r-1, d, ir - j) h_j^{\Delta}.$$

Surprisingly, the transformation described in Theorem 8 also appears in completely unrelated contexts. In Sect. 7, we will learn about one context. It seems to have first been studied by Holte [65] using a different description of the coefficients. It also appears in recent work of Diaconis and Fulman [44] (see also [45]) where different occurrences of the transformation are listed and their relations are studied.

5.2.3 Interval Subdivision

The transformation $\mathfrak{h}^{\Delta} \mapsto \mathfrak{h}^{\mathrm{Int}(\Delta)}$ is not well studied. Since the f-vector transformation is linear, it follows that also the h-vector transformation is linear. Nevertheless, there is no description of the coefficients known.

Example 9 The following are the matrices of the transformation $\mathfrak{h}^{\Delta} \mapsto \mathfrak{h}^{\text{Int}(\Delta)}$ for dim $\Delta = 1$ and dim $\Delta = 2$ respectively.

$$\begin{pmatrix} 1 & 0 & 0 \\ 3 & 4 & 3 \\ 0 & 0 & 1 \end{pmatrix} \qquad \begin{pmatrix} 1 & 0 & 0 & 0 \\ 16 & 14 & 10 & 7 \\ 7 & 10 & 14 & 16 \\ 0 & 0 & 0 & 1 \end{pmatrix}$$

The data suggests that the transformation is again non-negative and exhibits properties similar to the ones seen before. In particular, this suggests the following problem. Let $h_i^{\text{Int}(\Delta)} = \sum_{j=0}^{d} d_{i,j} h_j^{\text{Int}(\Delta)}$.

Problem 10 Are the numbers $d_{i,j}$ non-negative? If so, give a combinatorial interpretation of theses numbers.

Again, in the situation of the subdivision from [32, Fig. 1.2] the same problem can be posed. The $d_{i,j}$ will then be replaced by polynomials in N that specialize to $d_{i,j}$ for $N = 1$.

We have already seen that if Sub is one of our subdivision operations then the transformation $F_{\text{Sub}} : \mathfrak{f}^{\Delta} \mapsto \mathfrak{f}^{\text{Sub}(\Delta)}$ is a linear map—indeed invertible. The transformation $FH : \mathfrak{f}^{\Delta} \mapsto \mathfrak{h}^{\Delta}$ is also an invertible linear transformation. Thus it follows that also the map $\mathfrak{h}^{\Delta} \mapsto \mathfrak{h}^{\text{Sub}(\Delta)}$ is an invertible linear transformation. Indeed, $H_{\text{Sub}} = FH \cdot H_{\text{Sub}} \cdot FH^{-1}$ and hence H_{Sub} and F_{Sub} are similar. In particular, they share the same eigenvalues and F_{Sub} is diagonalizable if and only H_{Sub} is. In particular, this shows that for three subdivision operations we are studying the h-vector transformation is diagonalizable. Barycentric and edgewise subdivision satisfy the additional property that the matrix of the H_{Sub} has non-negative entries. This immediately implies:

Corollary 11 *Let Δ be a $(d-1)$-dimensional simplicial complex for which $h_i^{\Delta} \geq 0$ for $0 \leq i \leq d$. If Sub is either a barycentric or an edgewise subdivision then, for $0 \leq i \leq d$,*

$$h_i^{\text{Sub}(\Delta)} \geq h_i^{\Delta}.$$

We will see in Sect. 6 that $h_i^{\Delta} \geq 0$ for $0 \leq i \leq d$ if Δ is Cohen-Macaulay over some field $\mathbb{K}$. In particular, the assumptions of Corollary 11 hold. In the Cohen-Macaulay case the conclusion of Corollary 11 is known to hold for general subdivision operations by Stanley [105, Theorem 4.10]. Note that in the formulation of [105, Theorem 4.10], the subdivision is assumed to be quasi-geometric. This assumption is actually weaker than our definition of subdivision. However, the results from [105] do not imply that the matrix of H_{Sub} has non-negative coefficients in general and that $h_i^{\text{Sub}(\Delta)} \geq h_i^{\Delta}$ for all Δ with $h_i^{\Delta} \geq 0$ for all i.

6 First Relations to Algebra

In this section we provide a first insight into how the subdivision operations effect algebraic properties of the Stanley-Reisner rings. First we note that if Sub is any subdivision operator then $|\Delta|$ and $|\mathrm{Sub}(\Delta)|$ are homeomorphic. Therefore, any algebraic property of $\mathbb{K}[\Delta]$ that only depends on the homeomorphism type of $|\Delta|$ will not be effected by a subdivision operation. There are quite a few properties that indeed depend on the homeomorphism type only. In order to study these properties we start by considering some geometric, homological and combinatorial conditions on Δ that are easily seen to be invariant under homeomorphisms. We say that a property of a simplicial complex is invariant under homeomorphisms if for two simplicial complexes Δ and Δ' where $|\Delta|$ and $|\Delta|$ are homeomorphic either both or none has this property.

Dimension Clearly, the dimension of a simplicial complex is invariant under homeomorphisms.

Pure A $(d-1)$-dimensional simplicial complex Δ is called pure if all its facets are of the same dimension. Simple geometric considerations show that purity is preserved under homeomorphisms.

Homology One of the basic fact from algebraic topology states that the (reduced) simplicial homology $\widetilde{H}_i(\Delta; \mathbb{K})$ with coefficients in $\mathbb{K}$ only depends on the homeomorphism type of Δ (see e.g. [81, Chap. 2]). Indeed, it is an equally standard fact that it only depends on the topology of $|\Delta|$ up to homotopy equivalence.

Euler-characteristic The reduced Euler-characteristic $\widetilde{\chi}(\Delta)$ of a simplicial complex Δ is the alternating sum

$$\widetilde{\chi}(\Delta) = \sum_{i=-1}^{d-1} (-1)^i f_i^{\Delta}.$$

It is a direct consequence of the definition of simplicial homology that

$$\widetilde{\chi}(\Delta) = \sum_{i\geq 0} \dim_{\mathbb{K}} \widetilde{H}_i(\Delta; \mathbb{K}).$$

In particular, it follows that the Euler-characteristic is invariant under homeomorphism.

f- and h-Vector Very simple examples or our formulas for the behavior of f- and h-vector transformations for barycentric or edgewise subdivision show that we cannot expect any invariance of the full f- or h-vector under homeomorphism. But $f_{-1}^{\Delta} = h_0^{\Delta} = 1$ and

$$h_d^{\Delta} = (-1)^{d-1} \sum_{i=-1}^{d-1} (-1)^i f_i^{\Delta} = (-1)^{d-1} \widetilde{\chi}(\Delta)$$

show that, for $(d-1)$-dimensional simplicial complexes, these entries and linear combinations of the f- and h-vector are invariant under homeomorphisms.

Now with these facts in our mind, we turn to algebraic properties of $\mathbb{K}[\Delta]$.

Dimension By $\dim \mathbb{K}[\Delta] = \dim \Delta + 1$, the invariance of Krull dimension of $\mathbb{K}[\Delta]$ under homeomorphisms follows immediately from the invariance of the dimension of Δ.

Depth The depth $\mathrm{depth}(A)$ of the standard graded $\mathbb{K}$-algebra A is the maximal length ℓ of a sequence $u_1, \ldots, u_\ell$ of distinct homogeneous elements of A such that u_{i+1} is a non-zerodivisor on $A/(u_1, \ldots, u_i)$ for $0 \le i \le \ell - 1$. A maximal sequence with these properties is also called a regular sequence for A. Munkres [82, Theorem 3.1] showed the not at all obvious fact that the depth of $\mathbb{K}[\Delta]$ is invariant under homeomorphisms.

Cohen-Macaulay The standard graded $\mathbb{K}$-algebra A is called Cohen-Macaulay if $\dim(A) = \mathrm{depth}(A)$. A simplicial complex Δ is called Cohen-Macaulay over the field $\mathbb{K}$ if $\mathbb{K}[\Delta]$ is a Cohen-Macaulay $\mathbb{K}$-algebra. Given that we already know that dimension and depth are invariant under homeomorphisms it follows that, for a fixed field $\mathbb{K}$, the Cohen-Macaulay property of Δ, resp. $\mathbb{K}[\Delta]$, is invariant under homeomorphisms. Now a simple calculation shows that if for a standard graded $\mathbb{K}$-algebra A, a homogeneous element $u \in A_1$ is a non-zerodivisor on A, then

$$\mathrm{Hilb}_A(t) = \frac{\mathrm{Hilb}_{A/(u)}(t)}{1-t}.$$

Since a 0-dimensional $\mathbb{K}$-algebra is a $\mathbb{K}$-algebra that is finite dimensional as a vector space, it then follows by induction that the h-polynomial $h_A(t)$ of a Cohen-Macaulay standard graded $\mathbb{K}$-algebra A has only non-negative coefficients. This confirms that the assumptions in Corollary 11 are fulfilled for Cohen-Macaulay simplicial complexes Δ.

The following result by Reisner [97, Theorem 1] shows that the property of being Cohen-Macaulay over a field $\mathbb{K}$ only depends on the characteristic of $\mathbb{K}$. For the formulation of the result we need to define the link of a face F in a simplicial complex Δ over some ground set Ω. For $F \in \Delta$ the link $\mathrm{lk}_\Delta(F)$ of F is the simplicial complex

$$\{ F' \subset \Omega : F \cap F' = \emptyset;\ F \cup F' \in \Delta \}.$$

Theorem 12 ([97, Theorem 1]) *A simplicial complex Δ is Cohen-Macaulay over the field $\mathbb{K}$ if and only if*

$$\widetilde{H}_i(\mathrm{lk}_\Delta(F); \mathbb{K}) = 0$$

for all $-1 \le i < \dim(\mathrm{lk}_\Delta(F))$ and all $F \in \Delta$.

By the universal coefficient theorem it then follows that Cohen-Macaulayness over $\mathbb{K}$ just depends on the characteristic of $\mathbb{K}$. In addition, the theorem easily implies that if Δ is a Cohen-Macaulay simplicial complex over some $\mathbb{K}$ then it is pure.

Betti Numbers If $A = S/I$ for the polynomial ring $S = \mathbb{K}[x_1, \ldots, x_n]$ and I a homogeneous ideal in S then we can consider free resolutions of A as a module over S. A free resolution of A over S is an exact complex

$$\mathcal{F} : \cdots \xrightarrow{\partial_{i+1}} S^{\beta_i} \xrightarrow{\partial_i} S^{\beta_{i-1}} \to \cdots \xrightarrow{\partial_1} S^{\beta_0} \xrightarrow{\partial_0} A \to 0.$$

A resolution $\mathcal{F}$ is called minimal if $\partial_i(S^{\beta_i}) \subseteq \mathfrak{m} S^{\beta_{i-1}}$ for the maximal graded ideal $\mathfrak{m} = (x_1, \ldots, x_n)$ of S. It is well known that for minimal free resolutions the numbers β_i are determined by A. Indeed, it can be shown that a free resolution is minimal if and only if all β_i simultaneously have the minimal possible value over all free resolutions. Moreover, there is a minimal free resolution for which the differentials ∂_i are homomorphisms of graded modules after the standard grading on the free modules S^{β_i} is suitably twisted. For this let $S(-j)$ be the graded free S-module of rank 1 where the degree of each homogeneous element is shifted by j. If

$$\mathcal{F} : \cdots \xrightarrow{\partial_{i+1}} \bigoplus_{j>0} S(-j)^{\beta_{i,j}} \xrightarrow{\partial_i} \bigoplus_{j>0} S(-j)^{\beta_{i-1,j}} \to \cdots \xrightarrow{\partial_1} \bigoplus_{j\geq 0} S(-j)^{\beta_{0,j}} \xrightarrow{\partial_0} A \to 0$$

is a minimal graded free resolution then the $\beta_{i,j}$ are again determined by A and $\beta_i = \sum_{j\geq 0} \beta_{i,j}$. Clearly, the numbers β_i and $\beta_{i,j}$ depend on A. Therefore, we write $\beta_i(A)$ and $\beta_{i,j}(A)$ if it is ambiguous to which $\mathbb{K}$-algebra the Betti numbers belong to. We refer the reader to [91] for more details on free resolutions.

For $A = \mathbb{K}[\Delta]$ a formula by Hochster describes the graded Betti numbers of A in terms of some homological invariants of Δ. Hochster's Formula (see for example [29, Theorem 5.5.1]) reads as follows:

$$\beta_{i,i+j}(\mathbb{K}[\Delta]) = \sum_{\substack{W \subseteq [n] \\ \#W = i+j}} \dim_{\mathbb{K}} \widetilde{H}_{j-1}(\Delta_W; \mathbb{K}) \qquad (5)$$

where $\Delta_W = \{F \in \Delta : F \subseteq W\}$. In particular, if Δ is a simplicial complex on vertex set Ω,

$$\beta_{i,i+j}(\mathbb{K}[\Delta]) \neq 0 \quad \Leftrightarrow \quad \exists\, W \subseteq \Omega,\ \#W = i+j \text{ such that } \widetilde{H}_{j-1}(\Delta_W; \mathbb{K}) \neq 0.$$

The projective dimension pdim(A) of a standard graded $\mathbb{K}$-algebra A is the largest i for which $\beta_i \neq 0$. By the Auslander-Buchsbaum formula, one knows that the projective dimension of a standard graded $\mathbb{K}$-algebra $A = S/I$ is $n - \text{depth}(A)$, where $S = \mathbb{K}[x_1, \ldots, x_n]$. Since depth is invariant under homeomorphisms, the projective dimension increases with the number of variables. But all our subdivision operations increase the number of vertices, which is the same as the number of variables. An elementary calculation of $f_0^{\mathrm{sd}(\Delta)}, f_0^{\Delta^{\langle r \rangle}}$ and $f_0^{\mathrm{Int}(\Delta)}$ then yields to the following:

$$\mathrm{pdim}(\mathrm{sd}(\Delta)) = \sum_{i \geq 0} f_i^{\Delta} - \mathrm{depth}(\mathbb{K}[\Delta]),$$

$$\mathrm{pdim}(\Delta^{\langle r \rangle}) = \binom{r+n-1}{n-1} - \mathrm{depth}(\mathbb{K}[\Delta]),$$

$$\mathrm{pdim}(\mathrm{Int}(\Delta)) = \sum_{i \geq 0} (2^{i+1} - 1) f_i^{\Delta} - \mathrm{depth}(\mathbb{K}[\Delta]).$$

Indeed not only the projective dimension changes but also the Betti numbers change quite drastically.

Example 13 We compute the Betti numbers in some simple examples. The Stanley-Reisner ring of $\mathbb{K}[\Delta_2]$ is the polynomial ring in 3 variables and hence $\beta_{0,0} = 1$ and all other $\beta_{i,j}$ are 0. For $\mathbb{K}[\mathrm{sd}(\Delta_2)]$ we obtain the following table of number $\beta_{i,i+j}$. Note that $\beta_{i,\ell} = 0$ for $\ell < i$.

```
i1 : loadPackage("SimplicialComplexes");

i2 : S:=QQ[x1,x2,x3,x12,x13,x23,x123];

i4 : f:=simplicialComplex{x1*x12*x123,x1*x13*x123,x2*x12*x123,
                          x2*x23*x123,x3*x13*x123,x3*x23*x123}

o4 = | x3x23x123 x2x23x123 x3x13x123 x1x13x123 x2x12x123
       x1x12x123 |

o4 : SimplicialComplex

i7 : i:=monomialIdeal(\,f)

o7 = monomialIdeal (x1*x2,  x1*x3,  x2*x3,  x3*x12,  x2*x13,
                    x12*x13, x1*x23, x12*x23, x13*x23)

o7 : MonomialIdeal of QQ[x1, x2, x3, x12, x13, x23, x123]

i8 : betti res i

            0 1  2 3 4
o8 = total: 1 9 16 9 1
        0: 1 .  . . .
        1: . 9 16 9 .
        2: . .  . . 1

o8 : BettiTally
```

We already reached the limits of our computers when trying to compute the graded Betti number of $\mathbb{K}[\mathrm{sd}^2(\Delta_2)]$.

The Betti numbers $\beta_{1,1+j}$, $j \geq 0$, of $\mathbb{K}[\Delta]$ encode the degrees of generators if the Stanley-Reisner ideal of Δ. Note, as a monomial ideal the ideal I_Δ has a unique minimal set of monomial generators. More precisely, $\beta_{i,i+1}$ is the number of minimal monomial generators of I_Δ of degree $j + 1$; equivalently it is the number of minimal non-faces of Δ of size $j + 1$. From the definition of barycentric and interval subdivision it is clear that $\mathrm{sd}(\Delta)$ and $\mathrm{Int}(\Delta)$ have only minimal non-faces of size 2. Simplicial complexes with this property are called flag. Note that in both cases the simplices are defined as chains in some partial order; the inclusion order on the $\Delta \setminus \{\emptyset\}$ for the barycentric subdivision and the inclusion order on intervals in $\Delta \setminus \{\emptyset\}$ for the interval subdivision. Form that it follows that the minimal non-faces consist of two element subsets of the partial order that are incomparable. For the edgewise subdivision the situation is different. By construction the minimal non-faces of the edgewise subdivision of a simplex are of size two. Recall, that by definition the simplices of $\Delta_{d-1}^{(r)}$ are subsets of the integer points in $r|\Delta_{d-1}|$ which satisfy a compatibility relation on its two element subsets. The following example shows that for a general simplicial complex Δ there can be minimal non-faces of size > 2 in its edgewise subdivision.

Example 14 Here we start with a simplicial complex that is a cone over the boundary of a triangle. On the left in Fig. 2 there is a geometric realization of this complex, where the cone point is in the barycenter of the triangle. On the right hand side the local picture around the barycenter in the rth edgewise subdivision is shown. The original simplicial complex has a minimal non-face of size three and this non-face is copied in the edgewise subdivision for each r.

Note that not all minimal non-face of Δ are "copied" to the edgewise subdivision. The situation is more subtle.

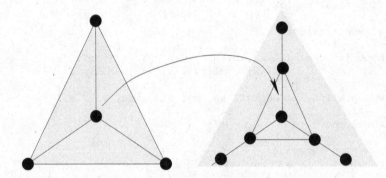

Fig. 2 A simplicial complex and the local picture in its rth edgewise subdivision

Gorenstein A standard graded $\mathbb{K}$-algebra A is Gorenstein if it is Cohen-Macaulay and $\beta_{\text{pdim}(A)} = 1$. Again, a simplicial complex Δ is called Gorenstein over the field $\mathbb{K}$ if $\mathbb{K}[\Delta]$ is a standard graded Gorenstein $\mathbb{K}$-algebra. We have seen already that Cohen-Macaulayness is invariant under homeomorphisms, but the condition on the last Betti number is not invariant under homeomorphisms. One should not be mislead by the example given in Example 13, where the 2-simplex and its barycentric subdivision are both Gorenstein. For example the second barycentric subdivision of the 1-simplex which is a path of length 4 on 5 vertices is a counterexample as shown below.

```
i1 : loadPackage("SimplicialComplexes");

i2 : S:=QQ[x1,x2,x3,x4,x5];

o2 : PolynomialRing

i3 : path := simplicialComplex{x1*x2,x2*x3,x3*x4,x4*x5}

o3 = | x4x5 x3x4 x2x3 x1x2 |

o3 : SimplicialComplex

i4 : betti res monomialIdeal(path)

            0 1 2 3
 o7 = total: 1 6 8 3
         0: 1 . . .
         1: . 6 8 3
```

Gorenstein* In order to avoid the introduction of additional algebraic notions that will not be used in the rest of the paper, we define the Gorenstein* property in homological terms. We refer to [29] or [107] for more details on the algebraic background. A simplicial complex Δ is called Gorenstein* if for all $F \in \Delta$ we have

$$\widetilde{H}_i(\text{lk}_\Delta(F); \mathbb{K}) = \begin{cases} 0 & \text{for } i \neq \dim \text{lk}_\Delta(F), \\ \mathbb{K} & \text{for } i = \dim \text{lk}_\Delta(F). \end{cases}$$

Thus Gorenstein* simplicial complexes can be seen as homology spheres. In particular, all triangulations of spheres are Gorenstein* simplicial complexes. It can be shown (Björner, Private communication) that being Gorenstein* is invariant under homeomorphisms. If Δ is Gorenstein* then its h-vector satisfies the Dehn-Sommerville equations [103]:

$$h_i^\Delta = h_{d-i}^\Delta \text{ for } 0 \leq i \leq d. \tag{6}$$

Gorenstein simplicial complexes are characterized as simplicial complexes that are the join of a full simplex and a Gorenstein* simplicial complex. Recall, that the join of two simplicial complexes Δ, Δ' on disjoint ground sets Ω, Ω' is the simplicial complex $\Delta * \Delta'$ on ground set $\Omega \cup \Omega'$ with faces $F \cup F'$ for $F \in \Delta$ and $F' \in \Delta'$. The preceding characterization of Gorenstein simplicial complexes explains why the first barycentric subdivision of a $(d-1)$-simplex is Gorenstein for all d. Indeed $\mathrm{sd}(\Delta_{d-1})$ is a cone over the barycenter of Δ_{d-1} and the barycentric subdivision of the boundary of Δ_{d-1}. Since the latter is a triangulation of a $(d-1)$-sphere Gorensteinness of $\mathrm{sd}(\Delta_{d-1})$ follows.

7 Few Subdivisions, Real Rootedness, Koszul and Veronese Algebras, Charney-Davis Conjecture

For a subdivision operator Sub denote by $\mathrm{Sub}^r(\Delta)$ the result of its r^{fold} application to Δ. In this section we want to study combinatorial and algebraic properties of $\mathrm{Sub}^r(\Delta)$ that are achieved for $r \gg 0$. Explicit bounds on R such that the property holds for $r \geq R$ are available in some cases but are left as open questions in many others. When an explicit bound R is available then $R = 1$. For interval subdivision the lack of knowledge on the f- and h-vector transformation implies that we can only speculate about the limiting behavior. We will ask some explicit questions and provide some examples.

7.1 Barycentric Subdivision

For $\Delta = \Delta_{d-1}$ the $(d-1)$-simplex Theorem 7 shows that $h_i^{\mathrm{sd}(\Delta_{d-1})} = A_{d+1,i}$ where $A_{d+1,i}$ is the Eulerian number counting the permutations in S_{d+1} with i descents. In particular,

$$\mathfrak{h}^{\mathrm{sd}(\Delta_{d-1})}(t) = \sum_{i=0}^{d} A_{d+1,d-i}t^i \tag{7}$$

is the Eulerian polynomial (see [108, Chap. 1.6]). A result due to Frobenius states that the Eulerian polynomial has only roots which are real, simple and negative.

The following result can be seen as a generalization of the theorem by Frobenius.

Theorem 15 ([25, Theorem 2]) *Let Δ be a $(d-1)$-dimensional simplicial complex such that $h_i^{\Delta} \geq 0$ then $\mathfrak{h}^{\mathrm{sd}(\Delta)|}(t)$ has only real, simple and negative roots.*

Note that the assertion that the roots are negative is a simple consequence of the assumption $h_i^{\Delta} \geq 0$ and Corollary 11. Since $\mathfrak{h}^{\Delta}(t)$ is real rooted if and only if $f^{\Delta}(t)$ is real rooted, any result on real rootedness for $\mathfrak{h}^{\Delta}(t)$ immediately transfers to $f^{\Delta}(t)$ and vice versa. The proof of Theorem 15 heavily relies on a result by Brändén [23, Theorem 4.1], which we state here in a variation geared toward the application in our setting.

Theorem 16 ([23, Theorem 4.1]) *Let $h_i \geq 0$, $0 \leq i \leq d$, and assume that*

$$\sum_{i\geq 0}^{d} h_i t^i (x+1)^{d-i} = \sum_{i=0}^{d} k_i \binom{t}{i}.$$

Then $\sum_{i=0}^{d} k_i t^i$ has only simple and real roots.

Indeed a finite number of subdivisions suffice for any simplicial complex.

Theorem 17 ([25, Theorem 3 (i)]) *Let Δ be a $(d-1)$-dimensional simplicial complex. There is a number $R > 0$ such that for $r \geq R$, the polynomial $\mathfrak{h}^{sd^r(\Delta)}(t)$ has only real roots.*

There is no good bound on R that can be obtained from the proof of [25, Theorem 3 (i)].

Problem 18 Give a bound R depending on some invariants of the simplicial complex Δ such that for $r \geq R$ the polynomial $\mathfrak{h}^{sd^r(\Delta)}(t)$ has only real roots.

From the point of view of combinatorics real rootedness is interesting by the following classical result of Newton (see [33, p. 270]).

Theorem 19 (Newton) *Let $h_d + h_{d-1}t + \cdots + h_0 t^d$ be a polynomial with non-negative coefficients that has only real roots. Then $h_i^2 \geq h_{i-1}h_{i+1}$ for $1 \leq i \leq d-1$.*

The property $h_i^2 \geq h_{i-1}h_{i+1}$ for $1 \leq i \leq d-1$ satisfied by the sequence $(h_0, \ldots, h_d)$ is called log-concavity. We say that $(h_0, \ldots, h_d)$ has no internal zeros if there is no index $0 < i < d$ for which $h_i = 0$ and $h_{i-1}, h_{i+1} \neq 0$. If $(h_0, \ldots, h_d)$ is non-negative and has no internal zeros then it is a very simple exercise to show that log-concavity implies unimodality, that is, there is some $0 \leq i \leq d$ such that

$$h_0 \leq \cdots \leq h_{i-1} \leq h_i \geq h_{i+1} \geq \cdots \geq h_d.$$

Thus we get as corollaries the following results.

Corollary 20 *Let Δ be a $(d-1)$-dimensional simplicial complex such that $h_i^\Delta \geq 0$ then $(h_0^{sd(\Delta)}, \ldots, h_d^{sd(\Delta)})$ is log-concave and if there are no internal zeros then it is unimodal.*

Since we know that $\mathfrak{h}^{sd(\Delta)}(t)$ is real rooted if and only if $\mathfrak{f}^{sd(\Delta)}(t)$ is real rooted, the consequences of this and subsequent corollaries also hold for the f-vector of $sd(\Delta)$. Note that the assumption $h_i^\Delta \geq 0$, $0 \leq i \leq d$, is satisfied for all complexes that are Cohen-Macaulay over some field $\mathbb{K}$.

A closer analysis of the h-vector transformation shows that for large r all entries $h_i^{\Delta^{(r)}}$ for $0 \leq i \leq d-1$ are positive. Thus we get:

Corollary 21 Let Δ be a $(d-1)$-dimensional simplicial complex. There is a number $R > 0$ such that for $r \geq R$ the sequence $(h_0^{\mathrm{sd}^r(\Delta)}, \ldots, h_d^{\mathrm{sd}^r(\Delta)})$ is log-concave and unimodal.

Next we provided a link between real rootedness of $\mathfrak{h}^{\mathrm{sd}(\Delta)}(t)$ and algebraic properties of $\mathbb{K}[\Delta]$. This will require some preparation. We first recall the concept of a Koszul algebra. Note that for a standard graded $\mathbb{K}$-algebra $A = \bigoplus_{i \geq 0} A_i$ by $A_0 = \mathbb{K} \cong A/\bigoplus_{i \geq 1} A_i$ the field $\mathbb{K}$ is an A-module. A standard graded $\mathbb{K}$-algebra A is called Koszul, if $\mathbb{K}$ has a linear resolution over A. That is there is an exact complex of graded A modules:

$$\mathcal{F} : \cdots \xrightarrow{\partial_{i+1}} A(-i)^{\alpha_i} \xrightarrow{\partial_i} A(i-1)^{\alpha_{i-1}} \to \cdots \xrightarrow{\partial_1} A(-0)^{\alpha_0} \xrightarrow{\partial_0} \mathbb{K}.$$

As usual by $A(-i)$ we denote the free A-module of rank 1 where the degrees of homogeneous elements are shifted by j. Equivalently, using homological notation one can say that A is Koszul if $\mathrm{Tor}_i^A(\mathbb{K}, \mathbb{K})_j = 0$ for $i \neq j$. By the exactness of $\mathcal{F}$ it follows that

$$\sum_{i \geq 0} \alpha_i t^i = \frac{1}{\mathrm{Hilb}_A(-t)}.$$

In particular, $\frac{1}{\mathrm{Hilb}_A(-t)}$ has non-negative coefficients in its Taylor expansion. In this situation Pringsheim's Theorem [114, Sect. 7.2] implies that, whenever $\mathrm{Hilb}_A(-t)$ has any (complex) zeros then $\sum_{i \geq 0} \alpha_i t^i$ has a pole at $-\rho$, the maximal modulus of a (complex) zero of $\mathrm{Hilb}_A(-t)$. In particular, $\mathrm{Hilb}_A(-\rho) = 0$ which implies $h_A(-\rho) = 0$ and therefore $h_A(t)$ has a real root. A necessary condition for $A = S/I$, $S = \mathbb{K}[x_1, \ldots, x_n]$ and I a homogeneous ideal, to be Koszul is that I must be generated in degree 2. Recall the following theorem of Fröberg:

Theorem 22 ([56]) If I is a monomial ideal in the polynomial ring $S = \mathbb{K}[x_1, \ldots, x_n]$ then S/I is Koszul if and only of I is generated by quadratic monomials.

Note that Fröberg proved this theorem also for non-commutative polynomial rings, which is the really hard part of his work.

By Fröberg [56], $\mathbb{K}[\Delta]$ is Koszul if and only if all minimal non-faces of Δ are of size two, or equivalently Δ is flag. Hence $\mathbb{K}[\mathrm{sd}(\Delta)]$ is Koszul for any simplicial complex Δ. Therefore, $\mathfrak{h}^{\mathrm{sd}(\Delta)}(t)$ has one real root forced by the above reasoning and $\mathbb{K}[\mathrm{sd}(\Delta)]$ being Koszul. Of course if the degree of $\mathfrak{h}^{\mathrm{sd}(\Delta)}(t)$ is odd then we do not need this complicated argument. Nevertheless, there is a more subtle relationship between Koszulness and real rootedness in general. This relationship has been revealed and worked out in [93]. In the sequel we give a brief overview of the relations detailed in [93] and refer the interested reader to [93] for further information.

As usual let $A = \bigoplus_{i \geq 0} A_i$ be a standard graded $\mathbb{K}$-algebra and $\mathfrak{m} = \bigoplus_{i \geq 1} A_i$ its unique graded maximal ideal. Set

$$F_i := A \otimes_{\mathbb{K}} \underbrace{\mathfrak{m} \otimes_{\mathbb{K}} \cdots \otimes_{\mathbb{K}} \mathfrak{m}}_{i}$$

and define $\partial_i : F_i \to F_{i-1}$ by

$$\partial_i([m_1| \cdots |m_i]) := m_1[m_2| \cdots |m_i]$$

$$+ \sum_{j=1}^{i-1} (-1)^j [m_1| \cdots |m_j m_{j+1}| \cdots |m_i]$$

$$+ (-1)^i m_i[m_1| \cdots |m_{i-1}].$$

Then

$$\mathcal{F}_B : \cdots \xrightarrow{\partial_2} F_1 \xrightarrow{\partial_1} F_0 \xrightarrow{\partial_0} \mathbb{K}$$

is a free resolution of $\mathbb{K}$ over A. It is called the normalized Bar-resolution. It is far from being minimal but it is an explicit resolution of the A-module $\mathbb{K}$ and an arbitrary standard graded $\mathbb{K}$-algebra A. In particular, we have $H_i(\mathcal{F}_B \otimes \mathbb{K}) = \operatorname{Tor}_i^A(\mathbb{K}, \mathbb{K})$. Next we show how to decompose $\mathcal{F}_B$ and hence $\mathcal{F}_B \otimes \mathbb{K}$ into graded pieces making $\mathcal{F}_B$ a free graded resolution of $\mathbb{K}$ over A. In particular, one obtains $H_i(\mathcal{F}_B \otimes \mathbb{K})_j = \operatorname{Tor}_i^A(\mathbb{K}, \mathbb{K})_j$. As a consequence, A is Koszul if and only if $H_i(\mathcal{F}_B \otimes \mathbb{K})_j = 0$ for $i \neq j$. In homological degree i of $\mathcal{F} \otimes \mathbb{K}$ a homogeneous element of degree ℓ is given as a linear combination of elements from $A_{j_1} \otimes \cdots \otimes A_{j_i}$ for compositions $\ell = j_1 + \cdots + j_i$ and $j_1, \ldots, j_i \geq 1$. Thus

$$(\mathcal{F} \otimes \mathbb{K})_{i,l} := \bigoplus_{j_1 + \cdots + j_i = \ell} A_{j_1} \otimes \cdots \otimes A_{j_i}.$$

Hence the strand of $\mathcal{F}_B \otimes \mathbb{K}$ in bidegree ℓ is given by the complex

$$(\mathcal{F}_B \otimes \mathbb{K})_\ell : 0 \to \mathcal{F}_{\ell,\ell} \to \cdots \to \mathcal{F}_{1,\ell} \to 0$$

and $H_i((\mathcal{F} \otimes \mathbb{K})_\ell) = \operatorname{Tor}_i^A(\mathbb{K}, \mathbb{K})_\ell$. Finally, this leads to the following characterization of the Koszul property:

$$A \text{ Koszul} \Leftrightarrow H_i((\mathcal{F} \otimes \mathbb{K})_\ell) = \operatorname{Tor}_i^A(\mathbb{K}, \mathbb{K})_\ell = 0, \text{ for } i \neq \ell. \tag{8}$$

Set $a_i = \dim_{\mathbb{K}} A_i$. The Koszul condition (8) yields:

$$a_1^\ell \geq \sum_{i=1}^{\ell-1} (-1)^{\ell+1-i} \sum_{j_1 + \cdots + j_i = \ell} a_{j_1} \cdots a_{j_i}.$$

This is the determinantal condition:

$$\det \begin{pmatrix} a_1 & a_2 & \cdots & a_{\ell-1} & a_\ell \\ 1 & a_1 & \cdots & a_{\ell-2} & a_{\ell-1} \\ 0 & 1 & \cdots & a_{\ell-3} & a_{\ell-2} \\ \cdot & \cdot & \cdots & \cdots & \cdot \\ \cdot & \cdot & \cdots & \cdots & \cdot \\ \cdot & \cdot & \cdots & \cdots & \cdot \\ 0 & 0 & \cdots & 1 & a_1 \end{pmatrix} \geq 0$$

Now the following theorem brings us back to real rooted polynomials.

Theorem 23 (Aissen et al. [4]) *Let $a_i \in \mathbb{R}_+$, $i \in \mathbb{N}$ such that*

$$\sum_{i \geq 0} a_i t^i = \frac{h(t)}{(1-t)^d} .$$

Then $h(t)$ has only real roots if and only if all minors of the infinite matrix

$$\begin{pmatrix} a_1 & a_2 & a_3 & \cdots & \cdots \\ 1 & a_1 & a_2 & \cdots & \cdots \\ 0 & 1 & a_1 & \cdots & \cdots \\ \cdot & \cdot & \cdot & \cdots & \cdots \\ \cdot & \cdot & \cdot & \cdots & \cdots \\ \cdot & \cdot & \cdot & \cdots & \cdots \end{pmatrix}$$

are ≥ 0.

In summary this shows that quite a few of the sufficient and necessary conditions for the real rootedness of $\mathfrak{h}^\Delta(t)$ are satisfied once the minimal non-faces of Δ are of size two. In [93] the connection is worked out in more detail and with more supportive evidence for a strong link. Nevertheless, a full understanding of the situation is still missing.

7.2 Edgewise Subdivision

For the edgewise subdivision the situation is less obvious.

Example 24 Let Δ be the boundary of the 5-simplex with h-vector equal to $(1, 1, 1, 1, 1, 1)$. Then

$$\mathfrak{h}^{\Delta^{(2)}}(t) = t^5 + 16 t^4 + 31 t^3 + 31 t^2 + 16 t + 1.$$

One easily checks that this polynomial has two non-real roots. For the 3rd edgewise subdivision the complex roots disappear and

$$\mathfrak{h}^{\Delta^{(3)}}(t) = t^5 + 51\,t^4 + 191\,t^3 + 191\,t^2 + 51\,t + 1$$

has only real roots. Nevertheless, there are more complicated simplicial complexes for which the h-polynomial of its rth edgewise subdivision has not only real roots also for some $r > 2$.

Indeed the following result shows that we can expect real rootedness for some r.

Theorem 25 ([26, Theorem 1.4]) *Let Δ be a $(d-1)$-dimensional simplicial complex. Then there is a number $R > 0$ such that for $r \geq R$ the polynomial $\mathfrak{h}^{\Delta^{(r)}}(t)$ has only real roots. Moreover, $\mathfrak{h}^{\Delta^{(r)}}(t)$ has strictly positive coefficients except possibly the constant coefficient.*

As for the barycentric subdivision, we get the following consequence.

Corollary 26 *Let Δ be a $(d-1)$-dimensional simplicial complex. There there is a number $R > 0$ such that for $r \geq R$ the sequence $(h_0^{\Delta^{(r)}}, \ldots, h_d^{\Delta^{(r)}})$ is log-concave and unimodal.*

The proof of Theorem 25 in [26] does not yield an explicit bound on R. We suspect that the following might hold.

Problem 27 Is it true that if Δ is a simplicial complex with $h_i^\Delta \geq 0$ for $0 \leq i \leq d$ then $\mathfrak{h}^{\Delta^{(r)}}(t)$ has only real roots for $r \geq d$?

Our discussion about Koszulness when studying barycentric subdivisions ties real rootedness to Koszulness and therefore to simplicial complexes that are flag. We know from Sect. 6 that there are complexes for which no edgewise subdivision has all minimal non-faces of size two. Thus looking for real rootedness of the h-polynomial for edgewise subdivision seems unmotivated. But indeed here again there is a strong connection to the Koszul property. It is provided by the Veronese construction on standard graded algebras and results by Backelin [9], Eisenbud et al. [53] and Brun and Römer [27]. In the following we explain this in detail. We start with a construction on general standard graded algebras that seems unrelated to our context.

For a standard graded $\mathbb{K}$-algebra $A = \bigoplus_{i \geq 0} A_i$ and a number $r \geq 1$ the rth Veronese algebra $A^{\langle r \rangle}$ is the standard graded algebra $\bigoplus_{i \geq 0} A_{ir}$ where elements from A_{ir} are given degree i and multiplication in inherited from A. If $S = \mathbb{K}[x_1, \ldots, x_n]$ and I is a homogeneous ideal in S then the rth Veronese algebra of $A = S/I$ can be considered as a quotient S_r/I_r, where

$$S_r = \mathbb{K}[y_{a_1 \cdots a_n} \ : \ a_1 + \cdots + a_n = r, a_1, \ldots, a_r \geq 0 \}$$

and I_r is the Kernel of the map $S_r \rightarrow A^{(r)}$ sending $y_{a_1 \cdots a_n}$ to $x_1^{a_1} \cdots x_n^{a_n}$. In order to connect Veronese algebras and edgewise subdivision we need to introduce some concepts from Gröbner basis theory. A term order in the polynomial ring $S = \mathbb{K}[x_1, \ldots, x_n]$ is a linear order $\preceq$ on the set of monomials Mon $= \{x_1^{a_1} \cdots x_n^{a_n} : a_1, \ldots, a_n \geq 0\}$ such that $1 \preceq m$ for all monomials m and $m \leq m'$ implies $mm'' \preceq m'm''$ for all monomials m, m', m''. For a polynomial $g \in S$ the leading monomial of g is the largest monomial with respect to $\preceq$ whose coefficient in the expansion of g as a linear combination of monomials is not zero. For an ideal I in S the initial ideal $\mathrm{in}_{\preceq}(I)$ of I with respect to $\preceq$ is the ideal generated by the leading monomials of the elements of I. Clearly, $\mathrm{in}_{\preceq}(I)$ is a monomial ideal. Using this terminology and the above notation, we can relate Veronese algebras of Stanley-Reisner rings and edgewise subdivisions.

Theorem 28 ([27, Corollary 6.5]) *For a $(d-1)$-dimensional simplical complex Δ on ground set $[n]$ there is a term order $\preceq$ such that for $\mathbb{K}[\Delta]^{(r)} = S_r/I_r$ we have $\mathrm{in}_{\preceq}(I_r) = I_{\Delta^{(r)}}$.*

Now by an elementary linear algebra argument it follows that for any homogeneous ideal I in S and any term order $\preceq$ we have

$$\mathrm{Hilb}_{S/I}(t) = \mathrm{Hilb}_{S/\,\mathrm{in}_{\preceq}(I)}(t). \tag{9}$$

Now (9) immediately implies the following.

Lemma 29 $h_{\mathbb{K}[\Delta]^{(r)}}(t) = h_{\mathbb{K}[\Delta^{(r)}]}(t)$.

In particular, Theorem 8 describes the transformation of the coefficients of the h-polynomial of $\mathbb{K}[\Delta]$ when taking the rth Veronese. Indeed, in [26] it is shown that Theorem 8 essentially describes the transformation of the h-polynomial of any standard graded $\mathbb{K}$-algebra A when taking the rth Veronese. The latter is true up to the some technical assumptions on the degree of the h-polynomial that will become irrelevant when taking high enough Veroneses.

Now the next result connects back to Koszulness.

Theorem 30 ([9, 53]) *Let A be a standard graded $\mathbb{K}$ algebra. Then there is an $R > 0$ such that $A^{(r)}$ is Koszul if $r \geq R$.*

With this result we can conclude:

Corollary 31 *Let Δ be a $(d-1)$-dimensional simplicial complex. Then there is a number R such that for $r \geq R$ the reciprocal h-polynomial $t^d \mathfrak{h}^{\Delta^{(r)}}(1/t)$ of the rth edgewise subdivision of Δ is the h-polynomial of a Koszul algebra.*

This now supports the suspicion that also for edgewise subdivisions we can expect real roots from some r on. We will see in Sect. 9 that this suspicion is justified.

7.3 Interval Subdivision

Since we do not have good control over the h-vector transformation for interval subdivisions we can only look at examples and speculate.

Example 32 Let Δ be a $(d-1)$-dimensional simplicial complex with $h_i^\Delta \geq 0$ for $0 \leq i \leq d$.

If $\dim \Delta = 2$, since $h_0^\Delta = 1$ it follows from Example 9 and a simple discriminant calculation that $\mathfrak{h}^{\mathrm{Int}(\Delta)}(t)$ is always real rooted.

For $\dim \Delta = 3$, with Example 9 and a little bit more effort, one can show that in our situation $\mathfrak{h}^{\mathrm{Int}(\Delta)}(t)$ is always real rooted.

Note, that we have already observed that the minimal non-faces of $\mathrm{Int}(\Delta)$ are all of size two. Hence $\mathrm{Int}(\Delta)$ is flag and therefore $\mathbb{K}[\mathrm{Int}(\Delta)]$ is Koszul.

Problem 33 Is $\mathfrak{h}^{\mathrm{Int}(\Delta)}(t)$ real rooted for all simplicial complexes Δ for which $h_i^\Delta \geq 0$ for $0 \leq i \leq \dim \Delta$?

7.4 Subdivisions and the Charney-Davis Conjecture

The above mentioned questions on real rootedness are relevant in the context of the Charney–Davis conjecture [31, Conjecture D]. This conjecture states that if Δ is a $(d-1)$-dimensional Gorenstein* simplicial complex over a field $\mathbb{K}$ whose minimal non-faces are of size two then

$$(-1)^{\lfloor \frac{d}{2} \rfloor} \cdot (h_0^\Delta - h_1^\Delta + \cdots \pm h_d^\Delta) \geq 0.$$

For d odd, the conjecture follows immediately from the Dehn-Sommerville equations (6) that hold for every Gorenstein* simplicial complex Δ. Nevertheless, there is an interesting version of the Charney-Davis conjecture also in this case; see [59]. For d even, the conjecture is a local and combinatorial version of a conjecture of Hopf on the Euler-characteristic of a certain class of manifolds; see [31].

We have already seen that subdividing Gorenstein* simplicial complexes leads to Gorenstein* simplicial complexes as Gorenstein* is invariant under homeomorphisms. In addition we know that $(d-1)$-dimensional Gorenstein* simplicial complexes satisfy the Dehn-Sommerville equations (6): $h_i^\Delta = h_{d-i}^\Delta$ for $0 \leq i \leq d$. This implies that if $\alpha \neq 0$ is a root of $\mathfrak{h}^\Delta(t)$ then also $1/\alpha$ is a root of $\mathfrak{h}^\Delta(t)$. Since the Charney-Davis conjecture is trivial for d odd we can assume that d is even. By $1 = h_0^\Delta = h_d^\Delta$ it follows that $\mathfrak{h}^\Delta(0) = h_d^\Delta = 1 \neq 0$ and therefore 0 is not a root. Hence we can write

$$\mathfrak{h}^\Delta(t) = \prod_{i=1}^{d/2} (t - \alpha_i)(t - 1/\alpha_i)$$

for some possibly non-zero complex numbers $\alpha_1, \ldots, \alpha_{d/2}$. Now assume that $\mathfrak{h}^{\Delta}(t)$ has only real roots. Since Gorenstein* implies Cohen-Macaulay we have that $h_i^{\Delta} \geq 0$ for $0 \leq i \leq d$. Together with $\mathfrak{h}^{\Delta}(0) = 1$ this implies that all roots of $\mathfrak{h}^{\Delta}(t)$ are strictly negative. In addition, they come in pairs, with one root less or equal to -1 and one root in $[-1, 0)$. Thus one of the factors $(-1 - \alpha_i)$ and $(-1 - 1/\alpha_i)$ is negative and one is positive. This shows that

$$\begin{aligned} (-1)^{d/2} &= \prod_{i=1}^{d/2}(-1 - \alpha_i)(-1 - 1/\alpha_i) \\ &= \mathfrak{h}^{\Delta}(-1) \\ &= h_d^{\Delta} - h_{d-1}^{\Delta} + \cdots + h_0^{\Delta} \\ &= h_0^{\Delta} - h_1^{\Delta} + \cdots + h_d^{\Delta} . \end{aligned}$$

In particular, one gets the following implication.

Proposition 34 *Let Δ be a $(d-1)$-dimensional Gorenstein* simplicial complex and assume that $\mathfrak{h}^{\Delta}(t)$ is real rooted. Then Δ satisfies the Charney-Davis conjecture. In particular, this holds for the barycentric subdivision of any Gorenstein* simplicial complex.*

Note that the conclusion of the lemma does not require the assumption that Δ is flag which is used in the Charney-Davis conjecture. Nevertheless, there is no reasonable class of Gorenstein* simplicial complexes known whose h-polynomial is real rooted and for which not all minimal non-faces are of size two. Also it should be mentioned that by a result of Gal [58] the h-polynomials of flag triangulations of spheres of dimension ≥ 5 in general are not real rooted. Proposition 34 is a weak version of [83, Theorem 1.1] where it is shown that the γ-vector (see [58]) of a barycentrically subdivided simplicial complex that satisfies (6) and has non-negative h-vector has a γ-vector that is the f-vector of a flag simplicial complex. The γ-vector is a transformation of the h-vector first studied in [58]. Since

$$(-1)^{\lfloor \frac{d}{2} \rfloor} \cdot (h_0^{\Delta} - h_1^{\Delta} + \cdots \pm h_d^{\Delta})$$

is one entry of the γ-vector Proposition 34 follows from their result.

For barycentric subdivisions of boundary complexes of simplicial polytopes the conclusion of Proposition 34 was already known by a result from [106]. But the results in [106] use methods completely different from real rootedness. Indeed the results from [106] hold for boundary complexes of arbitrary polytopes. For the boundary complex of an arbitrary polytope the barycentric subdivision can either be defined as the abstract simplicial complex whose ground set are the non-empty proper faces of the polytope and whose simplices are the chains of faces (see Fig. 3). Geometrically, it is the simplicial complex whose vertices are the barycenters of the non-empty proper faces and whose simplices are the convex hulls of the barycenters

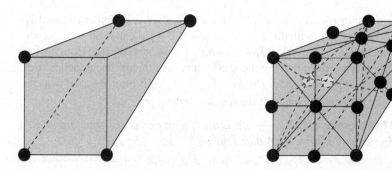

Fig. 3 Barycentric subdivision of the boundary complex of a non-simplicial polytope

corresponding to faces that form a chain. A positive answer to the following problem is suspected by Brenti and the second author.

Problem 35 . Let Δ be the barycentric subdivision of the boundary complex of an arbitrary polytope. Is $\mathfrak{h}^\Delta(t)$ real rooted ?

We note that if Δ is the barycentric subdivision of the boundary complex of the d-cube then $\mathfrak{h}^\Delta$ is the Eulerian polynomial of type B and hence real rooted by a result of Brenti [24]. Nevertheless, this does not immediately imply that barycentric subdivisions of cubical polytopes have real rooted h-polynomial. In particular, the question is open in this case and the class of cubical polytopes may indeed serve a good test case.

8 Few Subdivisions, Lefschetz Properties and Real Rootedness

In Sect. 7 we have seen how real rootedness of $\mathfrak{h}^\Delta(t)$ appears in the context of subdivisions. Its main combinatorial implication is the unimodality of the coefficient sequence of the h-vector in Corollaries 20, 21, and 26. One of the first and strongest results on unimodality of h-vectors is the g-theorem for boundary complexes of simplicial polytopes by Stanley [104] and Billera and Lee [20].

Theorem 36 (g-Theorem) *A sequence* $(h_0, \ldots, h_d) \in \mathbb{N}^{d+1}$ *is the h-vector of the boundary complex of a d-polytope if and only if*

(i) $h_0 = 1$ and $h_i = h_{d-i}$, $0 \le i \le d$.
(ii) There is a 0-dimensional standard graded $\mathbb{K}$-algebra $A = A_0 \oplus \cdots \oplus A_{\lfloor \frac{d}{2} \rfloor}$ for which $\dim_k A_i = h_i - h_{i-1}$, $1 \le i \le \lfloor \frac{d}{2} \rfloor$ (allowing $\dim A_i = 0$).

In particular, $(h_0, \ldots, h_d)$ is unimodal.

Note that a vector $(h_0, \ldots, h_s) \in \mathbb{N}^{s+1}$ for which there is a standard graded $\mathbb{K}$-algebra $A = A_0 \oplus \cdots \oplus A_s$ with $\dim_\mathbb{K} A_i = h_i$ for $0 \le i \le s$ is called an M-sequence. Clearly, in this situation $\text{Hilb}_A(t) = h_0 + \cdots + h_s t^s$. The name is derived from the fact that Macaulay characterized these sequences by a system of non-linear inequalities. Also note that the difference vector $(h_0, h_1 - h_1, \ldots, h_{\lfloor \frac{d}{2} \rfloor})$ is often referred to as the g-vector, which explains the name of the result.

Theorem 37 (Macaulay) $h_0 + \cdots + h_s t^s$ *is the Hilbert series of a standard graded graded* 0-*dimensional* $\mathbb{K}$-*algebra* A *if and only if*

- $h_0 = 1$,
- $\displaystyle \binom{a_{i+1}}{i} + \cdots + \binom{a_j}{j-1} \le h_{i-1}$ *for* $i \ge 2$ *where* $h_i = \binom{a_{i+1}}{i+1} + \binom{a_i}{i} + \cdots + \binom{a_j}{j}$

  for $a_{i+1} > \cdots > a_j \ge j \ge 1$.

The representation of a number n as $n = \binom{a_i}{i} + \binom{a_{i-1}}{i-1} + \cdots + \binom{a_j}{j}$ for $a_i > \cdots > a_j \ge j \ge 1$ is called an i-binomial representation and is well known to exist and to be unique.

The characterization of h-vectors of boundary complexes of simplicial polytopes from Theorem 36 had been conjectured by McMullen and indeed his conjecture extends to arbitrary simplicial spheres and Gorenstein* simplicial complexes. This conjecture is generally referred to as the g-conjecture for simplicial spheres and Gorenstein* simplicial complexes. This conjecture is still wide open (see [112] for a recent comprehensive survey on the state of the conjecture).

The main idea for the necessity part of the theorem by Stanley in [104] is to find for a boundary complex Δ of a simplicial polytope a standard graded $\mathbb{K}$-algebra $B = B_0 \oplus \cdots \oplus B_d$ such that $\dim_\mathbb{K} B_i = h_i^\Delta$ and an element $\omega \in B_1$ for which multiplication by ω^{d-2i} is an isomorphism of B_i and B_{d-i}. The latter implies that multiplication by ω is an injective map $A_i \xrightarrow{\omega} A_{i+1}$ for $0 \le i < \frac{d}{2}$ and surjective in the second half. In particular, $h_0 = \dim_\mathbb{K} B_0 \le \cdots \le h_{\lfloor \frac{d}{2} \rfloor} = \dim_\mathbb{K} B_{\lfloor \frac{d}{2} \rfloor}$ and $B/(\omega) \cong A_0 \oplus \cdots \oplus A_{\lfloor \frac{d}{2} \rfloor}$ where $\dim_\mathbb{K} A_i = h_i^\Delta - h_{i-1}^\Delta$ for $1 \le i \le \frac{d}{2}$ and $\dim_\mathbb{K} A_0 = 1$. Stanley in [104] uses toric geometry to identify the algebra $B = B_0 \oplus \cdots \oplus B_d$ as the even degree part of the cohomology algebra of the toric variety associated to the boundary complex Δ of the polytope. This algebra is known to be the quotient of the Stanley-Reisner ring of Δ by a regular sequence of linear forms for infinite fields $\mathbb{K}$. From this we infer that $\dim_\mathbb{K} B_i = h_i^\Delta$. Now Stanley invokes the hard Lefschetz theorem to find the element ω. This theorem has motivated several rivalling definitions of what is now called a Lefschetz algebra (see for example [22]).

Let us start with the definition of an s-Lefschetz element in a standard graded 0-dimensional $\mathbb{K}$-algebra $B = B_0 \oplus \cdots \oplus B_d$. Let s, $1 \le s \le d$, be a number. We call an element $\omega \in B_1$ an s-Lefschetz element if the multiplication map $\omega^{s-2i} : B_i \to B_{s-i}$ is injective for all $0 \le i \le \frac{s-1}{2}$. Now B is called a strong Lefschetz algebra if there is a d-Lefschetz element and B is called almost strong Lefschetz, if there is a $(d-1)$-Lefschetz element. Now the g-theorem uses in its proof the fact that there is a suitable quotient of $\mathbb{K}[\Delta]$ that is strong Lefschetz. Let ω be an s-Lefschetz element in $B = B_0 \oplus \cdots \oplus B_d$ and assume that for some $0 \le j \le \frac{s+1}{2}$ we have $\dim_{\mathbb{K}} B_0 \le \cdots \le \dim_{\mathbb{K}} B_j$ then the difference vector

$$\dim_{\mathbb{K}} B_0 + (\dim_{\mathbb{K}} B_1 - \dim_{\mathbb{K}} B_0)t + \cdots + (\dim_{\mathbb{K}} B_j - \dim_{\mathbb{K}} B_{j-1})t^j$$

is the Hilbert-series of the standard graded $\mathbb{K}$-algebra $B/(\omega, B_{j+1}, \ldots, B_d)$ and hence an M-sequence.

Next we study subdivision operations in the light of Lefschetz properties.

8.1 Barycentric Subdivision

The following result is the Lefschetz analog of Theorem 15.

Theorem 38 ([72, Theorem 1.1]) *Let Δ be a shellable $(d - 1)$-dimensional simplicial complex and $\mathbb{K}$ an infinite field. Then there is a regular sequence $\ell_1, \ldots, \ell_d$ of linear forms in $\mathbb{K}[\mathrm{sd}(\Delta)]$ such that $\mathbb{K}[\mathrm{sd}(\Delta)]/(\ell_1, \ldots, \ell_d)$ is almost strong Lefschetz.*

Recall that a simplicial complex is shellable if and only if it is pure and there is a linear order on its facets $F_1, \ldots, F_c$ such that for all $1 \le a < b \le c$ there is a number $a' < b$ such that $F_a \cap F_b \subseteq F_{a'} \cap F_b = F_b \setminus \{\omega\}$ for some $\omega \in F_b$. It is well known that shellability implies Cohen-Macaulayness (see [21] for more details and further references).

Now from Theorem 38 it follows that for $j \le \frac{d}{2}$ such that $h_0^{\mathrm{sd}(\Delta)} \le \cdots \le h_j^{\mathrm{sd}(\Delta)}$ the difference vector $(h_0, h_1^{\mathrm{sd}(\Delta)} - h_0^{\mathrm{sd}(\Delta)}, \ldots, h_j^{\mathrm{sd}(\Delta)} - h_{j-1}^{\mathrm{sd}(\Delta)}$ is an M-sequence. From Corollary 20 we know that for Δ with non-negative h-vector there is a j such that

$$h_0^{\mathrm{sd}(\Delta)} \le \cdots \le h_j^{\mathrm{sd}(\Delta)} \ge \cdots \ge h_d^{\mathrm{sd}(\Delta)}.$$

Let us call a j for which this holds a peak of $\mathfrak{h}^{\mathrm{sd}(\Delta)}$. Note, that the peak is not necessarily uniquely determined. Now Corollary 20 does not give an estimate for the peak in $\mathfrak{h}^{\mathrm{sd}(\Delta)}$.

Proposition 39 ([72, Corollary 4.7]) *If Δ is a $(d - 1)$-dimensional simplicial complex such that $h_i^{\Delta} \ge 0$ for $0 \le i \le d$. Then the peak of $\mathfrak{h}^{\mathrm{sd}(\Delta)}$ is at $\frac{d}{2}$ if d is even and at $\frac{d-1}{2}$ or at $\frac{d+1}{2}$ if d is odd.*

Before we formulate the combinatorial consequences of Theorem 38 we need the following result which allows us to formulate the consequences for Cohen-Macaulay simplicial complexes.

Theorem 40 ([107, Theorem 3.1]) *Let* $\mathfrak{h} = (h_0, \ldots, h_d) \in \mathbb{N}^{d+1}$ *be a sequence of non-negative integers. The following conditions are equivalent:*

(i) $\mathfrak{h}$ *is the h-vector of a shellable simplicial complex.*
(ii) $\mathfrak{h}$ *is the h-vector of a Cohen-Macaulay simplicial complex.*
(iii) $\mathfrak{h}$ *is an M-sequence.*

Now we are in position to provide the enumerative information coming from Theorem 38 in full detail.

Corollary 41 (Corollary 1.3   [72])   *Let* Δ *be a* $(d-1)$*-dimensional Cohen-Macaulay simplicial complex over some field* $\mathbb{K}$ *and let* j *be the peak of* $\mathfrak{h}^{\mathrm{sd}(\Delta)}$*. Then the vector*

$$(h_0, h_1^{\mathrm{sd}(\Delta)} - h_0^{\mathrm{sd}(\Delta)}, \ldots, h_j^{\Delta} - h_{j-1}^{\mathrm{sd}(\Delta)})$$

is M-sequence. In particular, the g-conjecture holds for barycentric subdivisions of simplicial spheres or Gorenstein* simplicial complexes.

Note that for Cohen-Macaulay simplicial complexes that do not satisfy the Dehn-Sommerville equations (6) Corollary 41 does not give any information about the second half of the h-vector.

Example 42 In the following we make an explicit computation in the situation of Theorem 38. Our simplicial complex Δ is the boundary of the octahedron on vertex set [6] (seen as a double cone over the quadrangle with vertices 2, 3, 4, 5) and with the face 256 removed. Clearly, Δ is shellable and therefore Theorem 38 applies.

```
i1 : loadPackage("SimplicialComplexes");

i2 : loadPackage("Depth");

i3 : S:=QQ[x1,x2,x3,x4,x5,x6,x12,x13,x14,x15,x23,x25,x62,x34,
           x63,x45,x64,x65,x123,x134,x145,x125,x623,x634,x645];

i4 : l:= {x1*x12*x123,x1*x13*x123,x2*x12*x123,x2*x23*x123,
                       x3*x13*x123,x3*x23*x123,
          x1*x13*x134,x1*x14*x134,x3*x13*x134,x3*x34*x134,
                       x4*x14*x134,x4*x34*x134,
          x1*x14*x145,x1*x15*x145,x4*x14*x145,x4*x45*x145,
                       x5*x15*x145,x5*x45*x145,
          x1*x15*x125,x1*x12*x125,x2*x12*x125,x2*x25*x125,
                       x5*x15*x125,x5*x25*x125,
          x6*x62*x623,x6*x63*x623,x2*x62*x623,x2*x23*x623,
                       x3*x63*x623,x3*x23*x623,
          x6*x63*x634,x6*x64*x634,x3*x63*x634,x3*x34*x634,
                       x4*x64*x634,x4*x34*x634,
          x6*x64*x645,x6*x65*x645,x4*x64*x645,x4*x45*x645,
                       x5*x65*x645,x5*x45*x645};

i5 : c:=simplicialComplex l;
```

```
i6 : i:=monomialIdeal(c);

i7 : reduceHilbert(hilbertSeries(S/i))

                    2
       1 + 22T + 19T
o7 = --------------
             3
         (1 - T)

o7 : Expression of class Divide

i8 : T:=S/i;

i9 : use T;
```

Now we construct a regular sequence.

```
i10 : rs:=matrix{{x1+x2+x3+x4+x5+x6,x12+x13+x14+x15+x23+x34+x45
        +x25+x62+x63+x64+x65,x123+x134+x145+x125+x623+x634+x645}};

i11 : B:=T/ideal(rs);

i13 : reduceHilbert(hilbertSeries(B))

                  2
       1 + 22T + 19T
o13 = --------------
             1

o13 : Expression of class Divide

i14 : use B;

i15 : omega:=x1+2*x2+3*x3+4*x4+5*x5+6*x6+x12+2*x13+3*x14+4*x15
        +5*x23+6*x34+7*x45+8*x25+9*x62+10*x63+11*x64+12*x65
        +x123+2*x134+3*x145+4*x125+5*x623+6*x634+7*x645;

i16 : reduceHilbert(hilbertSeries(B/ideal(omega)))

       1 + 21T
o16 = -------
          1

o16 : Expression of class Divide
```

Indeed, our element ω has the property that the multiplication map between two consecutive graded pieces of the algebra T has maximal possible rank. Hence, the parts from degree 3 on disappear when we mod out by ω. The remaining Hilbert-series is then as predicted by Theorem 38.

8.2 Edgewise Subdivision

Indeed there is no result known that relates Lefschetz properties and the Stanley-Reisner ring of edgewise subdivisions directly. But there is a result that makes the detour thorough Veronese algebras.

Theorem 43 ([71, Theorem 1.1]) *Let A be a standard graded Cohen-Macaulay $\mathbb{K}$-algebra of dimension d and let $\mathbb{K}$ be an infinite field. For a regular sequence $\ell_1, \ldots, \ell_d$ of elements of A_1 and $r \geq 1$ denote by $I_\ell^{(r)}$ the ideal generated by $\ell_1^r, \ldots, \ell_d^r$ in $A^{(r)}$. Then for $r \geq \deg h_A(t)$ the algebra $B = A^{(r)}/I_\ell^{(r)}$ is almost strong Lefschetz.*

In order to understand the combinatorial consequences of Theorem 43 it is necessary to recall from [72, Chap. 2] that under the assumption of Theorem 43 the elements $\ell_1^r, \ldots, \ell_d^r$ are a regular sequence of degree 1 elements in $A^{(r)}$. This shows that $B = A^{(r)}/I_\ell^{(r)}$ is a 0-dimensional standard graded $\mathbb{K}$-algebra whose Hilbert-series $\text{Hilb}_B(t)$ is the h-polynomial $h_A(t)$ of A.

Now consider the case when $A = \mathbb{K}[\Delta]$. The Cohen-Macaulay condition on A is by definition equivalent to Δ being Cohen-Macaulay over $\mathbb{K}$. Then for $B = \mathbb{K}[\Delta]^{(r)}/I_\ell^{(r)}$ it follows from Theorem 28 that

$$\text{Hilb}_B(t) = t^d h_{\mathbb{K}[\Delta^{(r)}]}(1/t).$$

The following result was obtained in [71] using the weak Lefschetz property of standard graded 0-dimensional $\mathbb{K}$-algebras. We do not want to introduce this additional concept here and refer the reader for example to [71] or [22]. In addition, the proof of the result uses some Gröbner basis arguments. We formulate the result from [71] only for the case $A = \mathbb{K}[\Delta]$.

Theorem 44 ([71, Theorem 1.3]) *Let Δ be a Cohen-Macaulay simplicial complex over the field $\mathbb{K}$. Then for $r \geq \deg \mathfrak{h}_{\mathbb{K}[\Delta]}(t)$*

(i) $\mathfrak{h}^{\Delta^{(r)}}$ *is the f-vector of a flag simplicial complex.*
(ii) $\mathfrak{h}^{\Delta^{(r)}}$ *is unimodal and if j is a peak then $(h_0^{\Delta^{(r)}}, h_1^{\Delta^{(r)}} - h_0^{\Delta^{(r)}}, \ldots, h_j^{\Delta^{(r)}} - h_{j-1}^{\Delta^{(r)}})$ is the f-vector of a simplicial complex.*

Note that the condition of being an f-vector is much stronger than the one being an M-sequence. By Gröbner basis theory a vector $(h_0, \ldots, h_d)$ is an M-sequence if and only if there is a set of monomials $\mathcal{M}$ such that if $m \in \mathcal{M}$ then all divisors of m are also in $\mathcal{M}$. Seeing faces of a simplicial complex as products over variables indexed by their element, the vector $(h_0, \ldots, h_d)$ is an f-vector if and only if there is a set $\mathcal{M}$ of squarefree monomials such that if $m \in \mathcal{M}$ then all divisors of m are in $\mathcal{M}$.

Thus for edgewise subdivisions we have obtained a connection to Lefschetz properties via the Veronese construction. Indeed the authors of [38] have been informed by Satoshi Murai that their results imply the following. There is no

$R > 0$ such that for $r \geq R$ there is a regular sequence $\ell_1^{\langle r \rangle}, \ldots, \ell_d^{\langle r \rangle}$ such that $\mathbb{K}[\Delta_{d-1}^{\langle r \rangle}]/(\ell_1^{\langle r \rangle}, \ldots, \ell_d^{\langle r \rangle})$ is (almost) strong Lefschetz. This leaves open the question of a subdivision operation that carries the Lefschetz property and connects to the Veronese construction.

Problem 45 Let Δ be a $(d-1)$-dimensional Cohen-Macaulay simplicial complex over some field $\mathbb{K}$. Is there a subdivision operation $\mathrm{Sub}(\bullet)$ satisfying the following two properties ?

- There is a term order $\preceq$ such that for $\mathbb{K}[\Delta]^{\langle r \rangle} = S_r/I_r$ we have $\mathrm{in}_{\preceq}(I_r) = I_{\mathrm{Sub}^r(\Delta)}$.
- There is an $R > 0$ such that for $r \geq R$ there is a regular sequence $\ell_1^{\langle r \rangle}, \ldots, \ell_d^{\langle r \rangle}$ such that $\mathbb{K}[\mathrm{Sub}^r(\Delta)]/(\ell_1^{\langle r \rangle}, \ldots, \ell_d^{\langle r \rangle})$ is (almost) strong Lefschetz.

8.3   Interval Subdivision

For interval subdivision the corresponding questions are wide open.

Problem 46 Let Δ be a $(d-1)$-dimensional Cohen-Macaulay simplicial complex over some field $\mathbb{K}$. Is there an $R > 0$ such that for $r \geq R$ there is a regular sequence $\ell_1^{\langle r \rangle}, \ldots, \ell_d^{\langle r \rangle}$ such that $\mathbb{K}[\mathrm{Int}^r(\Delta)]/(\ell_1^{\langle r \rangle}, \ldots, \ell_d^{\langle r \rangle})$ is (almost) strong Lefschetz ?

Since small examples do not give a good intuition for this problem we do not want to speculate about the answer.

8.4   Lefschetz Properties and Real Rootedness

By now we have two approaches to study the unimodality questions of h-vectors of subdivisions. One is real rootedness of the $\mathfrak{h}$-polynomial and the other is the (almost) strong Lefschetz property. The question arises if the two are related. Indeed there are results and conjectures that indicate that there is some relation.

The first result by Bell and Skandera [16] does not yet provide a connection but indicates the real rootedness is tied to M-vectors. We note that, the result in [16] actually works with a slightly weaker assumption.

Theorem 47 ([16, Theorem 3.6]) *Let* $h_d t^d + \cdots h_1 t + 1$ *be a real rooted polynomial with positive integer coefficients then* $(1, h_1, \ldots, h_d)$ *is an M-sequence.*

Indeed, in the paper the following problem is posed.

Problem 48 ([16, Question 1.1]) *Let* $f_{d-1} t^d + \cdots f_0 t + 1$ *be a real rooted polynomial with positive integer coefficients. Is* $(1, f_0, \ldots, f_{d-1})$ *the f-vector of a simplicial complex ?*

With a positive answer to this question Theorem 44 (i) (without the flagness and the explicit value of R) would already follow from Theorem 25.

We dare to go a bit beyond and ask the following question.

Problem 49 Let $h_d t^d + \cdots h_1 t + 1$ be a real rooted polynomial with positive integer coefficients and let j be a peak of $(1, h_1, \ldots, h_d)$. Is $(1, h_1 - h_0, \ldots, h_j - h_{j-1})$ an M-sequence or even an f-vector of a simplicial complex ?

With a positive answer to this question Theorem 44(ii) (without the explicit bounds) would already follow from Theorem 25. Note that this also ties in well with Problem 27 as the bounds on r given in Theorem 44 are all smaller than d.

9   Many Subdivisions, Limit Behaviour

In this section we want to study the behavior of invariants of $\mathbb{K}[\mathrm{Sub}^r(\Delta)]$ when r goes to infinity. We have already seen that some nice properties arise and stabilize for r sufficiently large. Some data that changes, such as the f- and h-vector, goes to infinity with r, so the question must be seen in the right light to make sense.

9.1   The f- and h-Vectors

9.1.1   Barycentric Subdivision

Theorem 50 ([25, Theorem 3]) *For a number $d \geq 1$ there are real negative numbers $\alpha_1, \ldots, \alpha_{d-2}$ such that for every $(d-1)$-dimensional simplicial complex Δ there are sequences $(\rho_i^{(r)})_{r \geq 1}$, $1 \leq i \leq d$, of complex numbers such that*

(i) $\rho_i^{(r)}$, $1 \leq i \leq d$, *are real for sufficiently large r.*
(ii) $\lim_{r \to \infty} \rho_i^{(r)} = \alpha_i$ *for $1 \leq i \leq d - 2$.*
(iii) $\lim_{r \to \infty} \rho_i^{(d-1)} = 0$.
(iv) $\lim_{r \to \infty} \rho_i^{(d)} = -\infty$.
(iv) *For all $r \geq 1$,*

$$\prod_{i=1}^{d}(t - \rho_i^{(r)}) = \mathfrak{h}^{\mathrm{sd}^r(\Delta)}(t).$$

Note that part (i) of the preceding theorem has appeared already as Theorem 17 in this article. The result is somewhat surprising, since shows that the limit behavior only depends on the dimension of Δ and not on the more subtle combinatorics of Δ. The following question immediately arises. Can one determine the limit roots $\alpha_1, \ldots, \alpha_{d-2}$. This question has been answered in [43, Theorems A and B], where a polynomial is exhibited whose roots are the limit roots and a transformation is given under which the set of roots is invariant.

9.1.2 Edgewise Subdivision

We start with a result identical to the one for barycentric subdivisions.

Theorem 51 ([26, Proposition 4.3]) *For a number $d \geq 1$ there are real negative numbers $\delta_1, \ldots, \delta_{d-2}$ such that for every $(d-1)$-dimensional simplicial complex Δ there are sequences $(\rho_i^{(r)})_{r \geq 1}$, $1 \leq i \leq d$, of complex numbers such that*

(i) $\rho_i^{(r)}$, $1 \leq i \leq d$, are real for sufficiently large r.

(ii) $\lim_{r \to \infty} \rho_i^{(r)} = \delta_i$ for $1 \leq i \leq d - 2$.

(iii) $\lim_{r \to \infty} \rho_i^{(d-1)} = 0$.

(iv) $\lim_{r \to \infty} \rho_i^{(d)} = -\infty$.

(iv) For all $r \geq 1$,

$$\prod_{i=1}^{d}(t - \rho_i^{(r)}) = \mathfrak{h}^{\langle r \rangle (\Delta)}(t).$$

A more thorough analysis of this theorem was performed in [15, Theorem 1.2]. They show some additional inequalities on the h-vector of $\Delta^{\langle r \rangle}$. In addition, they prove that the roots $\delta_1, \ldots, \delta_{d-1}$ from Theorem 51 are the roots of the Eulerian polynomial (7). In (7) we have already seen the Eulerian polynomial showing up in the context of subdivisions. But there they were linked to barycentric subdivisions. The following result indeed connects edgewise and barycentric subdivision but does not explain the above coincidence. For its formulation we denote for a face F of a simplicial complex Δ by $|F|$ the geometric realization of F as a subspace of the geometric realization of Δ. By $\partial |F|$ we denote the relative boundary of $|F|$ in its affine hull.

Proposition 52 ([38, Proposition 4.2]) *Let $r \geq d$ be positive integers. Then for any face F of $\Delta_{d-1}^{\langle r \rangle}$, satisfying*

$$\partial |F| = |F| \cap \partial |\Delta_{d-1}^{\langle r \rangle}|, \tag{10}$$

the link $\mathrm{lk}_{\Delta_{d-1}^{\langle r \rangle}}(F)$ is abstractly isomorphic to the barycentric subdivision of the boundary of a $(d - \#F)$-simplex.

9.1.3 Interval Subdivision

Having seen the two limiting results on iterated barycentric and edgewise subdivisions, one may wonder if there is something similar for interval subdivision. Indeed there is, except that it is not clear if the limiting roots are real. This is the content of following result from [43].

Theorem 53 ([43, Theorem 5.5]) *Let* Sub *be a subdivision operation on* $(d-1)$-*dimensional simplicial complexes that behaves identical on all simplices of the same dimension and such that for all* $(d-1)$-*dimensional simplicial complexes* Δ *we have* $f_{d-1}^{\mathrm{Sub}(\Delta)} > f_{d-1}^{\Delta}$. *Then there is a polynomial* $p_{\mathrm{Sub},d}(t)$ *of degree* $(d-2)$ *such that* $d-2$ *of the* d *complex roots of* $f^{\mathrm{Sub}^r(\Delta)}(t)$ *converge to the roots of* $p_{\mathrm{Sub},d}(t)$, *one root goes to zero and the other in absolute value to* ∞.

Clearly, most parts of the two limiting theorems on barycentric and edgewise subdivisions follow from this result. Indeed, a thorough analysis of the construction $p_{\mathrm{Sub},d}(t)$ from the proof of Theorem 53 would allow to deduce the result from [15] on the form of the limit roots in case of edgewise subdivisions.

Therefore, we are left with the following problem for interval subdivisions.

Problem 54 Describe the polynomial $p_{\mathrm{Sub},t}(t)$ from Theorem 53 where Sub is interval subdivision. Is it real rooted ?

9.2 Betti Numbers

Another set of invariants of a simplicial complex are the graded Betti numbers of $\mathbb{K}[\Delta]$. Again experiments suggest that under iterated subdivisions the nonzero Betti numbers go to infinity.

Here we can give the following non-vanishing result for barycentric and edgewise subdivisions. For its formulation we recall the Castelnuovo-Mumford regularity $\mathrm{reg}(A)$ of a standard graded $\mathbb{K}$-algebra A. The number $\mathrm{reg}(A)$ is the largest j for which there is a number i such that $\beta_{i,i+j} \neq 0$.

Theorem 55 ([38, Theorem 1.1]) *Let* Δ *be an arbitrary simplicial complex of dimension* $d - 1 > 0$. *Let* $\Delta(r)$ *be either* rth *barycentric subdivision* $\mathrm{sd}^r(\Delta)$ *or the* rth *edgewise subdivision* $\Delta^{\langle r \rangle}$ *of* Δ. *Then for large* r *the Castelnuovo-Mumford regularity of* $\mathbb{K}[\Delta(r)]$ *is given by:*

$$\mathrm{reg}(\mathbb{K}[\Delta(r)]) = \begin{cases} d - 1 & \text{if } \widetilde{H}_{d-1}(\Delta; \mathbb{K}) = 0, \\ d & \text{if } \widetilde{H}_{d-1}(\Delta; \mathbb{K}) \neq 0. \end{cases}$$

Furthermore:

(i) *For every* $j = 1, \ldots, d - 1$ *one has that* $\#\{i \,:\, \beta_{i,i+j}(\mathbb{K}[\Delta(r)]) = 0\}$ *is bounded above in terms of* d, j *(and independently of* r). *In particular:*

$$\lim_{r \to \infty} \frac{\#\{i \,:\, \beta_{i,i+j}(\mathbb{K}[\Delta(r)]) \neq 0\}}{\mathrm{pdim}(\mathbb{K}[\Delta(r)])} = 1.$$

(ii) *If* $\widetilde{H}_{d-1}(\Delta; \mathbb{K}) \neq 0$ *then*

$$\lim_{r \to \infty} \frac{\#\{i \,:\, \beta_{i,i+d}(\mathbb{K}[\Delta(r)]) \neq 0\}}{\mathrm{pdim}(\mathbb{K}[\Delta(r)])}$$

is a rational number in the interval $[0, 1)$ *that can be described in terms of the minimal* $(d - 1)$-*cycles of* Δ.

The paper [38] gives more precise bounds on the non-vanishing of the $\beta_{i,i+j}$ that differ for edgewise and barycentric subdivisions. But for both subdivision operations the result says that all strands $(\beta_{i,i+j'})_{i \geq 1}$ for $1 \leq j \leq j - 1$ are filled up almost from the beginning $i = 1$ up to the projective dimension. For the dth strand an interesting limiting behavior occurs. Surprisingly, the limit depends only on Δ and not on the chosen subdivision operation.

The results from Theorem 55(i) resemble results from [49, 50] and [119] on Betti numbers of high Veronese rings. Indeed, by Theorem 28 and the fact that Betti numbers weakly increase along flat deformations it follows that

$$\beta_{i,j}(\mathbb{K}[\Delta]^{\langle r \rangle}) \leq \beta_{i,j}(\mathbb{K}[\Delta^{\langle r \rangle}]).$$

Thus any non-vanishing result of from [49, 50] and [119] implies nonvanishing of the corresponding Betti numbers of the Stanley-Reisner ring of the edgewise subdivision of some Δ. Now the most general class of rings for which the result from [49, 50] and [119] hold, are Cohen-Macaulay rings. Therefore, if Δ is Cohen-Macaulay over $\mathbb{K}$ then the results from Theorem 55(i) are implied by the results from [49, 50] and [119]. On the other hand if Δ is not Cohen-Macaulay the non-vanishing results from [38] are not implied.

Problem 56

- Can one formulate conditions on a subdivision operation that are equivalent to the conclusions of Theorem 55(ii) ?
- Show that interval subdivisions satisfies the conclusions of Theorem 55(i).
- Are there subdivision operations that satisfy Theorem 55(i) but not Theorem 55(ii) ?
- Does the limit value for the dth strand have an algebraic significance ?

As mentioned in [38], the assumption on the subdivision operation needed to satisfy the conclusions of Theorem 55(i) are not very restrictive. Sufficient conditions can be easily extracted from the proofs in [38]. But exact geometric conditions equivalent to the phenomena described in Theorem 55 are not known. For interval subdivison the authors of [38] have checked sufficient conditions that imply the conclusions of Theorem 55(i), but the proof is not contained in the paper. The limit value in Theorem 55(ii) is a slightly more subtle issue. It remains rather mysterious how this strand behaves for more general subdivision operations.

10   Resolutions Supported on Subdivisions

In this section we study how subdivisions can be used to explicitly construct free resolutions of monomial ideals. For this we need to introduce the concept of a cellular resolution first studied in [14].

Let $S = \mathbb{K}[x_1, \ldots, x_n]$ and I a monomial ideal in S. Recall that a graded free resolution of $A = S/I$ is an exact sequence of graded S-modules

$$\mathcal{F} : \cdots \xrightarrow{\partial_{i+1}} \bigoplus_{j \geq 0} S(-j)^{\beta_{i,j}} \xrightarrow{\partial_i} \bigoplus_{j \geq 0} S(-j)^{\beta_{i-1,j}} \to \cdots \xrightarrow{\partial_1} \bigoplus_{j \geq 0} S(-j)^{\beta_{0,j}} \xrightarrow{\partial_0} A \to 0.$$

Now for a monomial ideal I the $\mathbb{K}$-algebra A is $\mathbb{N}^n$-graded where for $\mathbf{a} = (a_1, \ldots, a_n) \in \mathbb{N}^n$ the $\mathbf{a}$-graded piece $A_{\mathbf{a}}$ of A is the $\mathbb{K}$-vectorspaces spanned by the image of $x_1^{a_1} \cdots x_n^{a_n}$ in A. For the $\mathbb{N}^n$-graded S-module A one can study $\mathbb{N}^n$ graded free resolutions

$$\mathcal{F} : \cdots \xrightarrow{\partial_{i+1}} \bigoplus_{\mathbf{a} \in \mathbb{N}^n} S(-\mathbf{a})^{\beta_{i,\mathbf{a}}} \xrightarrow{\partial_i} \bigoplus_{\mathbf{a} \in \mathbb{N}^n} S(-\mathbf{a})^{\beta_{i-1,\mathbf{a}}} \to \cdots \xrightarrow{\partial_i} \bigoplus_{\mathbf{a} \in \mathbb{N}^n} S(-\mathbf{a})^{\beta_{0,\mathbf{a}}} \xrightarrow{\partial_0} A \to 0.$$

Here $S(-\mathbf{a})$ is $\mathbb{N}^n$ graded S-module where the monomial $x_1^{b_1} \cdots x_n^{b_n}$ is given degree $(a_1 + b_1, \ldots, a_n + b_n)$ and all differential are homomorphisms of $\mathbb{N}^n$-graded A-modules. Clearly, we can always arrange for $\beta_{0,\underline{0}} = 1$ and $\beta_{0,\mathbf{a}} = 0$ for $j \neq \underline{0}$.

This now allows for the definition of a cellular resolution of $A = S/I$ where I is a monomial ideal (see [13], [14]). Let Δ be a abstract simplicial complex with vertex set the set of exponent vectors $\mathbf{a} = (a_1, \ldots, a_n) \in \mathbb{N}^n$ of the monomials $x_1^{a_1} \cdots x_n^{a_n}$ from a not necessarily minimal monomial generating set of I. For $F \in \Delta$ we write $\mathbf{a}_F$ for the vector whose ith entry is componentwise maximum of the ith entries of the vectors in F. We use the convention $\mathbf{a}_\emptyset = \underline{0}$. Set $E_0 = S$ and

$$E_i = \bigoplus_{F \in \Delta, \dim F = i-1} S(-\mathbf{a}_F)$$

for $i \geq 1$ and choose of basis elements e_F of $S(-\mathbf{a}_F)$. We say that Δ supports a cellular resolution of $A = S/I$ if the $\mathbb{N}^n$-graded module homomorphisms $\partial_1(e_F) = 1$ and $\partial_i F_i \to F_{i-1}$ sending e_F to

$$\sum_{\mathbf{b} \in F} \varepsilon_{\underline{\mathbf{b}}} \underline{\mathbf{x}}^{\mathbf{a}_F - \mathbf{a}_{F-\mathbf{b}}} e_{F-\mathbf{b}}$$

where $\varepsilon_{\underline{\mathbf{b}}} = \pm 1$ is the sign from the simplicial differential and for $i \geq 2$ define a free resolution

$$\mathcal{F} : \cdots \xrightarrow{\partial_{i+1}} E_i \xrightarrow{\partial_i} E_{i-1} \xrightarrow{\partial_{i-1}} \cdots \xrightarrow{\partial_1} E_0 \xrightarrow{\partial_0} A \to 0.$$

Cellular resolutions have the advantage that algebraic resolutions can in some sense be given a global description, and they also lead to combinatorially interesting geometric complexes. Note that our definition is more restrictive then the one in [14] and tailored towards the applications in this section.

In [14] the authors give the following criterion, which characterizes the simplicial complexes that support a cellular resolution.

Proposition 57 ([14, Lemma 2.2]) *Let* $I = \langle \underline{\mathbf{x}}^{\mathbf{a}_1}, \ldots, \underline{\mathbf{x}}^{\mathbf{a}_r} \rangle$ *be a monomial ideal in* S *and* $\Omega = \{\mathbf{a}_1, \ldots, \mathbf{a}_r\}$. *A simplicial complex* Δ *over* Ω *supports a free resolution of* S/I *if and only if for all* $\underline{\mathbf{b}} \in \mathbb{N}^n$ *the simplicial complex* $\Delta_{\underline{\mathbf{b}}} = \{F \in \Delta : \mathbf{a}_F \leq \underline{\mathbf{b}}\}$ *is either* $\{\emptyset\}$ *or acyclic.*

Note that here we read $\mathbf{a}_F \leq \mathbf{a}_B$ as a componentwise inequality. Now let us see how subdivision can defined free resolutions.

10.1 Barycentric Subdivision

Let $I = \langle \underline{\mathbf{x}}^{\mathbf{a}_1}, \ldots, \underline{\mathbf{x}}^{\mathbf{a}_r} \rangle$ be a monomial ideal and $\Omega = \{\mathbf{a}_1, \ldots, \mathbf{a}_r\}$. By Proposition 57 the simplex $\Delta_{r-1} = 2^\Omega$ always supports a free resolution of I as $\Delta_{\underline{\mathbf{b}}} = 2^A$ where $A = \{\mathbf{a} \in \Omega : \mathbf{a} \leq \underline{\mathbf{b}}\}$ is a full simplex. This is a well known resolution and was first described in [113] without taking advantage of geometry. For that reason the resolution is called the Taylor resolution.

If in the barycentric subdivision $\mathrm{sd}(\Delta_{r-1})$ we label the face $F = \{A_1 \subset \cdots \subset A_s\}$ by $\mathbf{a}_{A_s}$ then we get $\mathrm{sd}(\Delta_{r-1})_{\underline{\mathbf{b}}} = \mathrm{sd}(2^A)$ for $A = \{\mathbf{a} \in \Omega : \mathbf{a} \leq \underline{\mathbf{b}}\}$ thus again by Proposition 57 it follows that $\mathrm{sd}(\Delta_{r-1})$ supports a resolution of $A = S/I$. Indeed with the same labeling it can be shown that if a simplicial complex Δ supports a cellular resolution of $A = S/I$ then so does $\mathrm{sd}(\Delta)$.

Next we want to recognize the barycentric subdivision as a subdivision induced by an arrangement of hyperplanes.

A hyperplane arrangement is a finite set of real (affine) hyperplanes is some $\mathbb{R}^d$. Our first example is the set A_{d-1} of hyperplanes

$$\{(x_1,\ldots,x_d) \,:\, x_i - x_j = 0\} \tag{11}$$

for all $1 \le i \le j \le d$. This arrangement is also known as the reflection arrangement for the symmetric group S_d. From this it also takes its name as it is the Coxeter arrangement of type A_{n-1}.

Lemma 58 *Let Γ_{d-1} be the convex hull of the unit basis vectors in $\mathbb{R}^d$. Then Γ_{d-1} is a geometric $(d-1)$-simplex and the hyperplane arrangement A_{d-1} cuts Γ_{d-1} into $d!$ simplices that are a geometric realization of the barycentric subdivision of the abstract $(d-1)$-simplex.*

10.2 Edgewise Subdivision

For the rth edgewise subdivision the natural labeling of the vertices as vectors $(a_1,\ldots,a_n)$ of non-negative integers summing up to r suggests that the edgewise subdivision might support a free resolution of the rth power $\mathfrak{m}^r$ of the maximal graded ideal $\mathfrak{m} = \langle x_1,\ldots,x_n \rangle$ in $S = \mathbb{K}[x_1,\ldots,x_n]$. Note, that a monomial $x_1^{a_1} \cdots x_n^{x_n}$ is among the minimal monomial generators of $\mathfrak{m}^r$ if and only if $a_1 + \cdots + a_n = r$.

Problem 59 Does the rth edgewise subdivision of the $(n-1)$-simplex Δ_{n-1} support a cellular resolution of $\mathfrak{m}^r$?

Figure 1b suggests that the edgewise subdivision of the standard simplex spanned by the unit basis vectors is achieved by intersecting the simplex with translates of the coordinate hyperplanes $x_i = 0$. Indeed, this is not correct as the following example shows (Fig. 4).

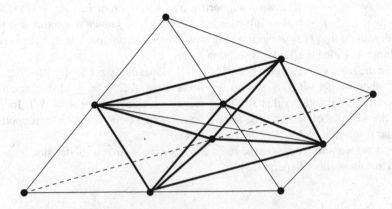

Fig. 4 Second edgewise subdivision of the 3-simplex

Example 60 We consider the 2nd edgewise subdivision of the standard 3-simplex dilated by 2.

The subdivision is induced by intersecting the 2-simplex with the hyperplanes defined by $x_i = 1$ for $1 \leq i \leq 4$ and by the hyperplane $x_1 + x_2 = 1$. The hyperplanes $x_i = 1$, $1 \leq i \leq 4$ cut out the octahedron drawn with bold edges. This octahedron is then triangulated into four simplices by the hyperplane $x_1 + x_2 = 1$.

Problem 61 Is the rth edgewise subdivision of the $(n - 1)$-simplex induced by intersecting the simplex with an arrangement of hyperplanes ?

There are well known minimal cellular resolutions of powers of the maximal idea (see [11, 75] and discussions therein), but they do not fit our picture here.

10.3 *Interval Subdivision*

Figure 1c shows that interval subdivision cannot be induced by intersecting the simplex with an arrangement of hyperplanes. In addition, it is not clear how interval subdivision naturally supports a cellular resolution of some monomial ideal. Nevertheless, we will see that a subdivision related to interval subdivision carries the cellular resolution of a monomial ideal and is induced by intersecting the simplex with an arrangement of hyperplanes.

Recall the reflection arrangement A_{n-1} of type A defined by hyperplanes $x_i - x_j = 0$ for $1 \leq i < j \leq n$. For our purposes it is of advantage to consider this arrangement as the graphic arrangement associated to the complete graph K_n. Here for a simple graph $G = ([n], E)$ on vertex set $[n]$ the corresponding graph arrangement is the arrangement of hyperplanes $x_i - x_j = 0$ for $\{i, j\} \in E$. We consider the restriction of the arrangement $\mathcal{H}_G$ to the subspace

$$U = \{x \in \mathbb{R}^n : x_n = 0, \, x_1 + x_2 + \cdots + x_{n-1} = 1\}, \qquad (12)$$

and we denote the restricted arrangement by $\mathcal{H}_G$. The bounded complex (i.e. the polyhedral complex consisting of bounded cells) of $\mathcal{H}_G$ will be denoted by $\mathcal{B}_G$. It is well-known that there is a one-to-one correspondence between the i-dimensional cells of $\mathcal{B}_G$, and the acyclic partial orientations of G with $i+2$ connected components having a unique source at n (see, e.g., [63, Corollary 7.3]). For the complete graph K_n the bounded complex $\mathcal{B}_G$ coincides with the barycentric subdivision of the $(n - 2)$-simplex.

Consider the polynomial ring $S = \mathbb{K}[x_1, \ldots, x_{n-1}]$ over the field $\mathbb{K}$ with variables corresponding to the vertices of G, and the polynomial ring $R = \mathbb{K}[y_{ij}, y_{ji} : \{i, j\} \in E(G)]$ with variables corresponding to the edges of G. For each vertex (0-dimensional face) p of $\mathcal{B}_G$ with corresponding partial orientation O_p we associate two monomials

$$\underline{x}(p) = \prod_{(i,j)\in O_p} x_j \quad \text{and} \quad \underline{y}(p) = \prod_{(i,j)\in O_p} y_{ij}$$

the multiplication being over all oriented edges of O_p. We define the graphic, and oriented graphic ideals of G as $\mathcal{M}_G = \langle \underline{x}(p) : p \in V(\mathcal{B}_G)\rangle$ and $\tilde{\mathcal{M}}_G = \langle \underline{y}(p) : p \in V(\mathcal{B}_G)\rangle$.

We now label each vertex p of the bounded complex $\mathcal{B}_G$ by the monomial $\underline{x}(p)$ (and similarly by $\underline{y}(p)$). By least common multiple construction we label all the higher dimensional faces of $\mathcal{B}_G$. Indeed the label of each face of $\mathcal{B}_G$ can be similarly obtained by taking an interior point f of that face with the associated partial orientation O_f and the corresponding monomial $\underline{x}(f) = \prod_{(i,j)\in O_f} x_j$ (and $\underline{y}(f) = \prod_{(i,j)\in O_f} y_{ij}$). The following result is an application of [85, Theorem 1.3(b)] for the graphic hyperplane arrangements.

Theorem 62 ([78, Theorem 7.2 and Corollary 10.7]) *The labeled polyhedral complex $\mathcal{B}_G$ supports a minimal free resolution for $\mathcal{M}_G$ and $\tilde{\mathcal{M}}_G$. In particular the Betti numbers are independent of the the characteristic of the base field, and the ith Betti number counts the number of i-dimensional regions of $\mathcal{B}_G$. Moreover $T(1, y)$ is the h-polynomial of $S/\mathcal{M}_G$, where $T(x, y)$ is the Tutte polynomial of the graph.*

Remark 63 The intersection lattice of $\mathcal{H}_G$ (i.e. the collection of nonempty intersections of hyperplanes ordered by reverse inclusion) is naturally isomorphic to the poset of connected 2-partitions of $V(G)$, see for example [63, p.112]. We recall that a region of an arrangement $\mathcal{H}$ is a connected component of the complement of the hyperplanes, i.e., $\mathbb{R}^n - \cup_{H\in\mathcal{H}} H$. Due to a result by Zaslavsky we know that the number of bounded regions of $\mathcal{H}$ can be obtained as an evaluation of its characteristic polynomial $|\chi_{\mathcal{H}}(1)|$. By Whitney's theorem the characteristic polynomial of an arrangement $\mathcal{H}$ (living in an n-dimensional vector space) is

$$\chi_{\mathcal{H}}(t) = \sum_{\substack{A\subseteq\mathcal{H} \\ A \text{ central}}} (-1)^{|A|} t^{n-\text{rank}(A)}, \tag{13}$$

where rank(A) denotes the rank of the $\cap_{h\in A} h$ in the intersection lattice of $\mathcal{H}$.

Here we extend some of the known results on graphic arrangements to the following class of arrangements: For all integers r and n, we denote by $\mathcal{H}_n^r$ the arrangement of hyperplanes in $\mathbb{R}^n$ given by

$$x_i - x_j = -r+1, -r+2, \ldots, r-1 \qquad \text{for all} \quad 1 \le i \le j \le n-1$$

and

$$x_i - x_n = 0 \qquad \text{for all} \quad 1 \le i \le n-1,$$

(a) 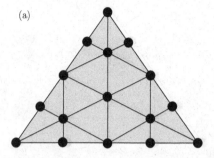 (b)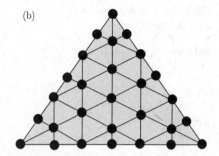

Fig. 5 The bounded complexes $\mathcal{B}_4^2$ (a) and $\mathcal{B}_4^3$ (b)

when we are restricting ourselves to the subspace U. We are only interested in the part of the arrangement living in the nonnegative side of the hyperplanes $\{x \in \mathbb{R}^n : x_i - x_n = 0\}$ for all i, i.e., those $x = (x_1, \ldots, x_n)$ such that $x_i - x_n \geq 0$. We denote the bounded complex (i.e. the polyhedral complex consisting of bounded cells) of $\mathcal{H}_n^r$ by $\mathcal{B}_n^r$.

Each face of $\mathcal{B}_n^r$ corresponds to an orientation of the complete graph K_n with multiple edges. The ideal $\mathcal{M}_n^r$ is defined in the polynomial ring $S = \mathbb{K}[x_1, \ldots, x_{n-1}]$ generated by monomials $\mathbf{x}(p) = \prod_{(i,j) \in O_p} x_j$, where the multiplication is over all oriented edges of O_p.

Example 64 Let $n = 4$, and consider the graph K_4. Then

$$\mathcal{M}_4 = \langle x_1^3, x_2^3, x_3^3, x_1^2 x_2^2, x_1^2 x_3^2, x_2^2 x_3^2, x_1 x_2 x_3 \rangle.$$

The bounded complex of $\mathcal{H}_4$ (which is the barycentric subdivision of the 2-simplex depicted in Fig. 1a supports the minimal free resolution of the ideal $\mathcal{M}_4$ by Postnikov and Shapiro [94, Theorem 6.1]. The bounded complex depicted in Fig. 5a supports the minimal free resolution of the ideal $\mathcal{M}_4^2$, where the parallel hyperplanes are corresponding to the hyperplanes $x_i - x_j = -1, 0, 1$ and the boundary hyperplanes are corresponding to $x_i - x_4 = 0$ for $1 \leq i < j \leq 3$. The bounded complex depicted in Fig. 5b supports the minimal free resolution of $\mathcal{M}_4^3$ and is cut out by the hyperplanes $x_i - x_j = -2, -1, 0, 1, 2$ and the boundary hyperplanes $x_i - x_4 = 0$ for $1 \leq i < j \leq 3$.

Theorem 65 ([76, Theorem 1]) *The bounded cell complex $\mathcal{B}_n^r$ supports a minimal free resolution for $\mathcal{M}_n^r$. In particular the ith Betti number counts the number of bounded regions of $\mathcal{B}_n^r$.*

For $n = 4$ and for all r, the ideal $\mathcal{M}_n^r$ coincides with the rth power of the ideal $\mathcal{M}_n$ which is not true in general (for $n > 4$).

We recall that given a polyhedral complex and a subset U of its vertices, its *induced subcomplex* on U, is the set of all its faces whose vertices belong to U. The polyhedral complex $\mathcal{B}_n$ associated to the ideal $\mathcal{M}_n$ naturally contains the subcomplexes induced on the vertices of top dimensional regions. In [76, Theorem 5.1] the

Fig. 6 The bounded complex
$\mathcal{B}_4^2$ with hyperplanes
$x_i - x_4 = c$ for $i = 1, 2, 3$
and some $c > 0$ (*dashed*)

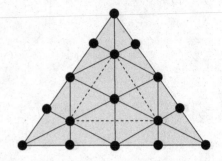

author shows that if we only take the generators of $\mathcal{M}_n$ corresponding to the subsets $A \subseteq T$ for a fixed subset $T \subset [n-1]$, then an induced subcomplex of $\mathcal{B}_n$ supports the minimal free resolution of the ideal generated by the monomials associated to the latter collection of subsets (see [77]).

The questions that immediately arise are:

Problem 66 Given an integer r and an arbitrary graph G, can we read the minimal free resolution of $\mathcal{M}_G^r$ as a deformation of the graphic hyperplane arrangement? How about the ideals whose generator set is a subset of $\mathcal{M}_G^r$?

If we add three additional hyperplanes (parallel to the hyperplanes $x_1 - x_4 = 0, x_2 - x_4 = 0$ and $x_3 - x_4 = 0$) to $\mathcal{H}_4^2$ then we obtain the same complex as the interval subdivision of the 2-simplex. See Figs. 6 and 1c.

Problem 67 Is there any relation between the interval subdivision of the $(n-2)$-simplex, and the bounded complex $\mathcal{B}_n^2$?

References

1. Abbott, J.: Quadratic interval refinement for real roots. ACM Commun. Comput. Algebra **48**, 3–12 (2014)
2. Abbott, J., Bigatti, A.M.: CoCoALib: a C++ library for doing Computations in Commutative Algebra. Available at http://cocoa.dima.unige.it/cocoalib
3. Abbott, J., Bigatti, A.M., Lagorio, G.: CoCoA-5: a system for doing Computations in Commutative Algebra. Available at http://cocoa.dima.unige.it/
4. Aissen, M., Schoenberg, I.J., Whitney, A.: On generating functions of totally positive sequences I. J. Anal. Math. **2**, 93–103 (1952)
5. Anick, D.: A counterexample to a conjecture of Serre. Ann. Math. **115**, 1–33 (1982)
6. Athanasiadis, C.A.: Edgewise subdivisions, local h-polynomials and excedances in the wreath product $\mathbb{Z}_r \wr \mathfrak{S}_n$. SIAM J. Disc. Math. **28**, 1479–1492 (2014)
7. Athanasiadis, C.A.: A survey of subdivisions and local h-vectors. In: The Mathematical Legacy of Richard P. Stanley, pp. 39–52. American Mathematical Society, Providence (2016)
8. Ay, N., Rauh, J.: Robustness and conditional independence ideals. Preprint (2011). arXiv:1110.1338
9. Backelin, J.: On the rates of growth of the homologies of Veronese subrings. In: Algebra, Algebraic Topology, and Their Interactions (Stockholm, 1983). Lecture Notes in Mathematics. vol. 1183, pp. 79–100, Springer, Berlin (1986)
10. Backelin, J., Fröberg, R.: Koszul algebras, Veronese subrings, and rings with linear resolutions. Rev. Roumaine Math. Pures Appl. **30**, 85–97 (1985)
11. Batzies, E., Welker, V.: Discrete Morse theory for cellular resolutions. J. Reine Angew. Math. **543**, 147–168 (2002)
12. Bayer, D., Stillman, M.: A criterion for detecting m-regularity. Invent. Math. **87**, 1–11 (1987)
13. Bayer, D., Sturmfels, B.: Cellular resolutions of monomial modules. J. Reine Angew. Math. **502**, 123–140 (1998)
14. Bayer, D., Peeva, I., Sturmfels, B.: Cellular resolutions of monomial modules. Math. Res. Lett. **5**, 31–46 (1998)
15. Beck, M., Stapledon, A.: On the log-concavity of Hilbert series of Veronese subrings and Ehrhart series. Math. Z. **264**, 195–207 (2010)
16. Bell, J., Skandera, M.: Multicomplexes and polynomials with real zeros, Discrete. Math. **307**, 668–682 (2007)
17. Bigatti, A.M., De Negri, E.: Stanley decompositions using *CoCoA*. In: Monomial Ideals, Computations and Applications. Lecture Notes in Mathematics, vol. 2083, pp. 47–59. Springer, Heidelberg (2013)
18. Bigatti, A.M., La Scala, R., Robbiano, L.: Computing toric ideals. J. Symbolic Comput. **27**, 351–365 (1999)

© Springer International Publishing AG 2017

A.M. Bigatti et al. (eds.), *Computations and Combinatorics in Commutative Algebra*, Lecture Notes in Mathematics 2176, DOI 10.1007/978-3-319-51319-5

123

19. Billera, L.J., Björner, A.: Face numbers of polytopes and complexes. In: Handbook of Discrete and Computational Geometry. CRC Press Online - Series: Discrete Mathematics and Its Applications, pp. 291–310. CRC, Boca Raton (1997)
20. Billera, L.J., Lee, C.W.: Sufficiency of McMullen's conditions for f-vectors of simplicial polytopes. Bull. Am. Math. Soc. New Series **2**, 181–185 (1980)
21. Björner, A.: Shellable and Cohen-Macaulay partially ordered sets. Trans. Am. Math. Soc. **260**, 159–183 (1980)
22. Boij, M., Migliore, J.C., Miró-Roig, R.M., Nagel, U., Zanello, F.: On the shape of a pure O-sequence. Mem. Am. Math. Soc. **218**(1024), viii+78 pp. (2012)
23. Brändén, P.: On linear transformations preserving the Pólya frequency property. Trans. Am. Math. Soc. **358**, 3697–3716 (2006)
24. Brenti, F.: q-Eulerian polynomials arising from Coxeter groups. Eur. J. Combin. **15**, 417–441 (1994)
25. Brenti, F., Welker, V.: f-vectors of barycentric subdivisions. Math. Z. **259**, 849–865 (2008)
26. Brenti, F., Welker, V.: The Veronese construction for formal power series and graded algebras. Adv. Appl. Math. **42**, 545–556 (2009)
27. Brun, M., Römer, T. : Subdivisions of toric complexes. J. Algebraic Combin. **21**, 423–448 (2005)
28. Bruns, W., Conca, A.: Gröbner bases and determinantal ideals. In: Commutative Algebra, Singularities and Computer Algebra (Sinaia, 2002). Nato Science Series II: - Mathematics, Physics and Chemistry, vol. 115, pp. 9–66. Kluwer Acadamic Publication, Dordrecht (2003)
29. Bruns, W., Herzog, J.: Cohen-Macaulay rings. In: Cambridge Studies in Advanced Mathematics, vol. 39. Cambridge University Press, Cambridge (1993)
30. Bruns, W., Vetter, U.: Determinantal Rings. Lecture Notes in Mathematics, vol. 1327. Springer, Berlin (1988)
31. Charney, R., Davis, M.: Euler characteristic of a nonpositively curved, piecewise Euclidean manifold. Pac. J. Math. **171**, 117–137 (1995)
32. Cheeger, J., Müller, W, Schrader, R.: On the curvature of piecewise flat spaces. Commun. Math. Phys. **92**, 405–454 (1984)
33. Comtet, L.: Advanced Combinatorics, Revised and Enlarged edition. D. Reidel, Dordrecht (1974)
34. Conca, A.: Koszul Algebras and their sygyzies. In: Combinatorial Algebraic Geometry. Lecture Notes in Mathematics, vol. 2108, pp. 47–59. Springer, Cham (2014)
35. Conca, A., Rossi, M.E., Valla, G.: Gröbner flags and Gorenstein algebras. Compos. Math. **129**, 95–121 (2001)
36. Conca, A., Trung, N.V., Valla, G.: Koszul property for points in projective spaces. Math. Scand. **89**, 201–216 (2001)
37. Conca, A., De Negri, E., Rossi, M.E.: Koszul Algebras and regularity. In: Commutative Algebra, pp. 285–315. Springer, New York (2013)
38. Conca, A., Kubitzke, M., Welker, V.: Asymptotic syzygies of Stanley-Reisner rings of iterated subdivisions. Trans. Am. Math. Soc. (2017, to appear)
39. Cox, D., Little, J., O'Shea, D.: Ideals, varieties, and algorithms. In: An Introduction to Computational Algebraic Geometry and Commutative Algebra. Undergraduate Texts in Mathematics, 4th edn. Springer, Cham (2015)
40. De Loera, J., Rambau, J., Santos, F.: Triangulations. Algorithms and Computation in Mathematics, vol. 25. Springer, Heidelberg (2010)
41. De Negri, E.: Toric rings generated by special stable sets of monomials. Math. Nachr. **203**, 31–45 (1999)
42. Decker, W., Greuel, G.-M., Pfister, G.: Primary decomposition: algorithms and comparisons. In: Algorithmic Algebra and Number Theory (Heidelberg, 1997), pp. 187–220. Springer, Berlin (1999)
43. Delucchi, E., Pixton, A., Sabalka, L.: Face vectors of subdivided simplicial complexes. Discret. Math. **312**, 248–257 (2012)

44. Diaconis, P., Fulman, J.: Carries, shuffling, and an amazing matrix. Am. Math. Mon. **116**, 788–803 (2009)
45. Diaconis, P., Fulman, J.: Carries, shuffling, and symmetric functions. Adv. Appl. Math. **43**, 176–196 (2009)
46. Diaconis, P., Sturmfels, B.: Algebraic algorithms for sampling from conditional distributions. Ann. Stat. **26**, 363–397 (1998)
47. Drton, M., Sturmfels, B., Sullivant, S.: Lectures on Algebraic Statistics. Oberwolfach Seminars, vol. 39. Birkhäuser, Basel (2009)
48. Edelsbrunner, H., Grayson, D.R.: Edgewise subdivision of a simplex. Discret. Comput. Geom. **24**, 707–719 (2000)
49. Ein, L., Lazarsfeld, R.: Asymptotic syzygies of algebraic varieties. Invent. Math. **190**, 603–646 (2012)
50. Ein, L., Erman, D., Lazarsfeld, R.: Asymptotics of random Betti tables. J. Reine Angew. Math. **702**, 55–75 (2015)
51. Eisenbud, D., Sturmfels, B.: Binomial ideals. Duke Math. J. **84**, 1–45 (1996)
52. Eisenbud, D., Huneke, C., Vasconcelos, W.: Direct methods for primary decomposition. Invent. Math. **110**, 207–235 (1992)
53. Eisenbud, D., Reeves, A., Totaro, B.: Initial ideals, Veronese subrings, and rates of algebras. Adv. Math. **109**, 168–187 (1994)
54. Fink, A.: The binomial ideal of the intersection axiom for conditional probabilities. J. Algebraic Combin. **33**, 455–463 (2011)
55. Freudenthal, H.: Simplizialzerlegung von beschränkter Flachheit. Ann. Math. **43**, 580–582 (1942)
56. Fröberg, R.: Determination of a class of Poincaré series. Math. Scand. **37**, 29–39 (1975)
57. Fröberg, R.: Koszul algebras. In: Advances in Commutative Ring Theory (Fez 1997). Lecture Notes in Pure and Applied Mathematics, vol. 205, pp. 337–350. Dekker, New York (1999)
58. Gal, S.R.: Real root conjecture fails for five and higher dimensional spheres. Discret. Comput. Geom. **34**, 269–284 (2005)
59. Gal, S.R., Januszkiewicz, T.: Odd-dimensional Charney-Davis conjecture. Discret. Comput. Geom. **44**, 802–804 (2010)
60. Gianni, P., Trager, B., Zacharias, G.: Gröbner bases and primary decompositions of polynomial ideals. J. Symbolic Comput. **6**, 149–167 (1988)
61. Grayson, D.R.: Exterior power operations in higher K-theory. K-Theory **3**, 247–260 (1989)
62. Grayson, D.R., Stillman, M.E.: Macaulay2, a software system for research in algebraic geometry. Available at http://www.math.uiuc.edu/Macaulay2/
63. Greene, C., Zaslavsky, T.: On the interpretation of Whitney numbers through arrangements of hyperplanes, zonotopes, non-Radon partitions, and orientations of graphs. Trans. Am. Math. Soc. **280**, 97–126 (1983)
64. Herzog, J., Hibi, T., Hreinsdottir, F., Kahle, T., Rauh, J.: Binomial edge ideals and conditional independence statements. Adv. Appl. Math. **45**, 317–333 (2010)
65. Holte, J.: Carries, combinatorics and an amazing matrix. Am. Math. Mon. **104**, 138–149 (1997)
66. Jensen, A.N.: Gfan, a software system for Gröbner fans and tropical varieties. Available at http://home.imf.au.dk/jensen/software/gfan/gfan.html
67. Jensen, A N.: CaTS, a software system for toric state polytopes. Available at http://www.soopadoopa.dk/anders/cats/cats.html
68. Kahle, T.: Decompositions of binomial ideals in Macaulay2. J. Softw. Algebra Geom. **4**, 1–5 (2012)
69. Kreuzer, M., Robbiano, L.: Computational Commutative Algebra 1. Springer, Berlin (2008). Corrected reprint of the 2000 original
70. Krick, T., Logar, A.: An algorithm for the computation of the radical of an ideal in the ring of polynomials. In: Applied Algebra, Algebraic Algorithms and Error-Correcting Codes (New Orleans, LA, 1991). Lecture Notes in Computer Science, vol. 539, pp. 195–205. Springer, Berlin (1991)

71. Kubitzke, M., Murai, S.: Lefschetz properties and the Veronese construction. Math. Res. Lett. **19**, 1043–1053 (2012)
72. Kubitzke, M., Nevo, E.: The Lefschetz property for barycentric subdivisions of shellable complexes. Trans. Am. Math. Soc. **361**, 6151–6163 (2009)
73. Laubenbacher, R.C., Swanson, I.: Permanental ideals. J. Symb. Comput. **30**, 195–205 (2000)
74. Lauritzen, S.L.: Graphical Models. Oxford Statistical Science Series, vol. 17. Oxford Science Publications/The Clarendon Press/Oxford University Press, New York (1996)
75. Mermin, J.: The Eliahou-Kervaire resolution is cellular. J. Commut. Algebra **2**, 55–78 (2010)
76. Mohammadi, F.: Deformation of hyperplane arrangements, characteristic polynomials, and resolutions of powers of ideals. In preparation (2014)
77. Mohammadi, F.: Divisors on graphs, orientations, syzygies, and system reliability. J. Algebraic Combin. **43**, 465–483 (2016)
78. Mohammadi, F., Shokrieh, F.: Divisors on graphs, binomial and monomial ideals, and cellular resolutions. Math. Z. **283**, 59–102 (2016)
79. Mora, T., Robbiano, L.: The Gröbner fan of an ideal. J. Symb. Comput. **6**, 183–208 (1988)
80. Morey, S., Villarreal, R.: Edge ideals: algebraic and combinatorial properties. Prog. Commut. Algebra **1**, 85–126 (2012)
81. Munkres, J.R.: Elements of Algebraic Topology. Addison-Wesley Publishing Company, Menlo Park (1984)
82. Munkres, J.R.: Topological results in combinatorics. Michigan Math. J. **31**, 113–128 (1984)
83. Nevo, E., Petersen, T.K., Tenner, B.E.: The γ-vector of a barycentric subdivision. J. Combin. Theory Ser. A **118**, 1364–1380 (2011)
84. Newman, M.E.J.: The structure and function of complex networks. SIAM Rev. **45**, 167–256 (2003)
85. Novik, I., Postnikov, A., Sturmfels, B.: Syzygies of oriented matroids. Duke Math. J. **111**, 287–317 (2002)
86. Ohtani, M.: Graphs and ideals generated by some 2-minors. Commun. Algebra **39**, 905–917 (2011)
87. Ojeda, I.: Binomial canonical decompositions of binomial ideals. Commun. Algebra **39**, 3722–3735 (2011)
88. Ojeda, I., Piedra-Sánchez, R.: Canonical decomposition of polynomial ideals. Unpublished preprint. http://departamento.us.es/da/prepubli/prepub54.pdf (1999)
89. Ojeda, I., Piedra-Sánchez, R.: Index of nilpotency of binomial ideals. J. Algebra **255**, 135–147 (2002)
90. Ortiz, V.: Sur une certaine decomposition canonique d'un idéal en intersection d'idéaux primaires dans un anneau noetherien commutatif. C. R. Acad. Sci. Paris **248**, 3385–3387 (1959)
91. Peeva, I.: Graded syzygies. Algebra and Applications, vol. 14. Springer, London (2011)
92. Pistone, G., Riccomagno, E., Wynn, H.P.: Algebraic Statistics. Computational Commutative Algebra in Statistics. Monographs on Statistics and Applied Probability, vol. 89. Chapman & Hall/CRC, Boca Raton (2001)
93. Polishchuk, A., Positselski, L.: Quadratic Algebras. University Lecture Series, vol. 37. American Mathematical Society, Providence (2005)
94. Postnikov, A., Shapiro, B.: Trees, parking functions, syzygies, and deformations of monomial ideals. Trans. Am. Math. Soc. **356**, 3109–3142 (2004)
95. Priddy, S.B.: Koszul resolutions. Trans. Am. Math. Soc. **152**, 39–60 (1970)
96. Reiner, V., Welker, V.: On the Charney–Davis and Neggers–Stanley conjectures. J. Comb. Theory Ser. A **109**, 247–280 (2005)
97. Reisner, G.A.: Cohen-Macaulay quotients of polynomial rings. Adv. Math. **21**, 30–49 (1976)
98. Roos, J.-E.: Commutative non Koszul algebras having a linear resolution of arbitrarily high order. Applications to torsion in loop space homology. C. R. Acad. Sci. **316**, 1123–1128 (1993)
99. Roos, J.-E.: Good and bad Koszul algebras and their Hochschild. J. Pure Appl. Algebra **201**, 295–327 (2005)

100. Sáenz-de-Cabezón, E., Wynn, H.P.: Measuring the robustness of a network using minimal vertex covers. Math. Comput. Simul. **104**, 82–94 (2014)
101. Savvidou, C.: Face numbers of cubical barycentric subdivisions. Preprint (2010). arXiv:1005.4156
102. Shimoyama, T., Yokoyama, K.: Localization and primary decomposition of polynomial ideals. J. Symbolic Comput. **22**, 247–277 (1996)
103. Sommerville, D.: The relations connecting the angle sums and volume of a polytope in space of n dimensions. Proc. R. Soc. Ser. A **115**, 103–119 (1927)
104. Stanley, R.P.: The number of faces of a simplicial convex polytope. Adv. Math. **35**, 236–238 (1980)
105. Stanley, R.P.: Subdivisions and local h-vectors. J. Am. Math. Soc. **5**, 805–851 (1992)
106. Stanley, R.P.: Flag f-vectors and the cd-index. Math. Z. **216**, 483–499 (1994)
107. Stanley, R.P.: Combinatorics and Commutative Algebra. Progress in Mathematics, 2nd edn, vol. 41. Birkhäuser, Boston (1996)
108. Stanley, R.P.: Enumerative Combinatorics, Volume 1. Cambridge Studies in Advanced Mathematics, 2nd edn, vol. 49. Cambridge University Press, Cambridge (2012)
109. Studený, M.: Attempts at axiomatic description of conditional independence. Kybernetika (Prague) Suppl. **25**(1–3), 72–79 (1989); Workshop on Uncertainty Processing in Expert Systems (Alšovice, 1988)
110. Sturmfels, B.: Gröbner Bases and Convex Polytopes. University Lecture Series, vol. 8. American Mathematical Society, Providence (1996)
111. Swanson, I., Taylor, A.: Minimal primes of ideals arising from conditional independence statements. J. Algebra **392**, 299–314 (2013)
112. Swartz, E.: Thirty five years and counting. Preprint (2014) arXiv:1411.0987
113. Taylor, D.: Ideals Generated by Monomials in an R-Sequence. Ph.D. Thesis, University of Chicago (1960)
114. Titchmarsh, E.C.: The Theory of Functions, 2nd edn. Oxford University Press, Oxford (1985)
115. Villarreal, R.: Monomial Algebras. Monographs and Textbooks in Pure and Applied Mathematics, vol. 238. Marcel Dekker, New York (2001)
116. von zur Gathen, J., Gerhard, J.: Modern Computer Algebra, 2nd edn. Cambridge University Press, Cambridge (2003)
117. Walker, J.W.: Canonical homeomorphisms of posets. Eur. J. Combin. **9**, 97–107 (1988)
118. Yao, Y.: Primary decomposition: compatibility, independence and linear growth. Proc. Am. Math. Soc. **130**, 1629–1637 (2002)
119. Zhou, X.: Effective non-vanishing of asymptotic adjoint syzygies. Proc. Am. Math. Soc. **142**, 2255–2264 (2014)

LECTURE NOTES IN MATHEMATICS 🐎 Springer

Editors in Chief: J.-M. Morel, B. Teissier;

Editorial Policy

1. Lecture Notes aim to report new developments in all areas of mathematics and their applications – quickly, informally and at a high level. Mathematical texts analysing new developments in modelling and numerical simulation are welcome.

   Manuscripts should be reasonably self-contained and rounded off. Thus they may, and often will, present not only results of the author but also related work by other people. They may be based on specialised lecture courses. Furthermore, the manuscripts should provide sufficient motivation, examples and applications. This clearly distinguishes Lecture Notes from journal articles or technical reports which normally are very concise. Articles intended for a journal but too long to be accepted by most journals, usually do not have this "lecture notes" character. For similar reasons it is unusual for doctoral theses to be accepted for the Lecture Notes series, though habilitation theses may be appropriate.

2. Besides monographs, multi-author manuscripts resulting from SUMMER SCHOOLS or similar INTENSIVE COURSES are welcome, provided their objective was held to present an active mathematical topic to an audience at the beginning or intermediate graduate level (a list of participants should be provided).

   The resulting manuscript should not be just a collection of course notes, but should require advance planning and coordination among the main lecturers. The subject matter should dictate the structure of the book. This structure should be motivated and explained in a scientific introduction, and the notation, references, index and formulation of results should be, if possible, unified by the editors. Each contribution should have an abstract and an introduction referring to the other contributions. In other words, more preparatory work must go into a multi-authored volume than simply assembling a disparate collection of papers, communicated at the event.

3. Manuscripts should be submitted either online at www.editorialmanager.com/lnm to Springer's mathematics editorial in Heidelberg, or electronically to one of the series editors. Authors should be aware that incomplete or insufficiently close-to-final manuscripts almost always result in longer refereeing times and nevertheless unclear referees' recommendations, making further refereeing of a final draft necessary. The strict minimum amount of material that will be considered should include a detailed outline describing the planned contents of each chapter, a bibliography and several sample chapters. Parallel submission of a manuscript to another publisher while under consideration for LNM is not acceptable and can lead to rejection.

4. In general, **monographs** will be sent out to at least 2 external referees for evaluation.

   A final decision to publish can be made only on the basis of the complete manuscript, however a refereeing process leading to a preliminary decision can be based on a pre-final or incomplete manuscript.

   Volume Editors of **multi-author works** are expected to arrange for the refereeing, to the usual scientific standards, of the individual contributions. If the resulting reports can be forwarded to the LNM Editorial Board, this is very helpful. If no reports are forwarded or if other questions remain unclear in respect of homogeneity etc, the series editors may wish to consult external referees for an overall evaluation of the volume.

5. Manuscripts should in general be submitted in English. Final manuscripts should contain at least 100 pages of mathematical text and should always include

   – a table of contents;
   – an informative introduction, with adequate motivation and perhaps some historical remarks: it should be accessible to a reader not intimately familiar with the topic treated;
   – a subject index: as a rule this is genuinely helpful for the reader.
   – For evaluation purposes, manuscripts should be submitted as pdf files.

6. Careful preparation of the manuscripts will help keep production time short besides ensuring satisfactory appearance of the finished book in print and online. After acceptance of the manuscript authors will be asked to prepare the final LaTeX source files (see LaTeX templates online: https://www.springer.com/gb/authors-editors/book-authors-editors/manuscriptpreparation/5636) plus the corresponding pdf- or zipped ps-file. The LaTeX source files are essential for producing the full-text online version of the book, see http://link.springer.com/bookseries/304 for the existing online volumes of LNM). The technical production of a Lecture Notes volume takes approximately 12 weeks. Additional instructions, if necessary, are available on request from lnm@springer.com.

7. Authors receive a total of 30 free copies of their volume and free access to their book on SpringerLink, but no royalties. They are entitled to a discount of 33.3 % on the price of Springer books purchased for their personal use, if ordering directly from Springer.

8. Commitment to publish is made by a *Publishing Agreement*; contributing authors of multiauthor books are requested to sign a *Consent to Publish form*. Springer-Verlag registers the copyright for each volume. Authors are free to reuse material contained in their LNM volumes in later publications: a brief written (or e-mail) request for formal permission is sufficient.

Addresses:
Professor Jean-Michel Morel, CMLA, École Normale Supérieure de Cachan, France
E-mail: moreljeanmichel@gmail.com

Professor Bernard Teissier, Equipe Géométrie et Dynamique,
Institut de Mathématiques de Jussieu – Paris Rive Gauche, Paris, France
E-mail: bernard.teissier@imj-prg.fr

Springer: Ute McCrory, Mathematics, Heidelberg, Germany,
E-mail: lnm@springer.com

Printed in the United States
By Bookmasters